T0135443

Junbo Gong

Spectral Kinetic Simulation of the Ideal Multipole Resonance Probe

Logos Verlag Berlin

λογος

Bibliographic information published by the Deutsche Nationalbibliothek

The Deutsche Nationalbibliothek lists this publication in the Deutsche
Nationalbibliografie; detailed bibliographic data are available
on the Internet at http://dnb.d-nb.de

ISBN 978-3-8325-5606-8

Logos Verlag Berlin GmbH
Georg-Knorr-Str. 4, Geb. 10,
D-12681 Berlin
Germany

Tel.:   +49 (0)30 / 42 85 10 90
Fax: +49 (0)30 / 42 85 10 92
http://www.logos-verlag.de

Spectral Kinetic Simulation
of the Ideal Multipole Resonance Probe

Written by

Junbo Gong

Born in Sichuan

Dissertation for the degree of Doctor of Engineering of the

Faculty of Electrical Engineering and Information Technology at the

Ruhr-Universität Bochum

2022

Supervisors:

1. Prof. Dr. rer. nat. Ralf Peter Brinkmann

2. Prof. Dr.-Ing. Peter Awakowicz

Submission date: June 24, 2022
Defense date: November 9, 2022

Acknowledgment

As I look back on my Ph.D. journey, I realize that I never even would be able to start it without the support and encouragement of others. I want to express my deepest gratitude to those who have helped me along the way.

First of all, I am deeply indebted to my supervisor Prof. Ralf Peter Brinkmann, who gave me an entirely new perception of physics. I want to thank him for his patience and kind encouragement. Each discussion with him helped me to vastly improve my grasp on the fundamentals. His mentorship significantly broadened my knowledge and understanding not only in academic matters but in life matters as well. I would also like to thank my supervisor Prof. Peter Awakowicz, and colleagues from the chair of General Electrical Engineering and Plasma Technology for the experimental work related to my project, which I have benefited tremendously from.

I am keenly aware of the debt I owe personally and professionally to Prof. Jens Oberrath and Michael Friedrichs, who I could always count on. Thank them for many fruitful discussions and generous support. I really enjoyed our collaborations, and I learned a lot from them. Thanks should also go to Dr. Michael Klute, Dr. Stefan Ries, and Dr. Felix Mitschker for being great colleagues and amazing friends. I enjoyed the pleasant moments spent with them over the years.

I am proud and honored to have been part of the chair of Theoretical Electrical Engineering. My colleagues have always been great. Many thanks to Dr. Liang Xu, Schabnam Naggary, Dr. Sebastian Wilczek, and Youfan He for sharing scientific ideas and providing kind assistance. In addition, I am grateful to Dennis Engel and Maximilian Klich for the exercises and exams we organized together. It was delightful to work with them. I also appreciate the help with financial and administrative issues from Ms. Elke Konhäuser and technical support from Mr. Ke Yan. Besides, special thanks to House of Plasma, and Dr. Moritz Oberberg for the experimental support.

I could not have undertaken this journey without tremendous emotional support and constant encouragement from my family and friends in China. Thousands of miles separate us, and yet my parents are always there for me, who both managed to give me the strength to keep on going and help me become the person I am today.

Lastly, immense gratitude goes to my girlfriend Romina Kestner and her family. Especially thank her for always believing in me no matter the circumstances and supporting me in everything I ventured.

Abstract

Active plasma resonance spectroscopy (APRS) is a process-compatible plasma diagnostic method that utilizes the natural ability of plasmas to resonate near the electron plasma frequency. The *Multipole Resonance Probe* (MRP) is a particular realization of APRS that has a high degree of geometric and electric symmetry. This radio-frequency driven probe of the spherical design is used for the supervision and control of low-temperature plasmas. The principle of the MRP can be described on the basis of an idealized geometry that is specifically suited for theoretical investigations.

Over the last decade, many studies of the MRP have been conducted to understand the resonance behavior of the plasma via the cold plasma model. However, in a pressure regime of a few Pa or lower, kinetic effects become important, which cannot be predicted by the cold plasma model. Therefore, in this work, a dynamic model of the interaction of the idealized MRP with a plasma is established, which is named the spectral kinetic model. Specifically, the self-consistent system is described in the Hamiltonian formalism, and the Poisson problem is solved explicitly with a Green's function. The proposed scheme reveals the kinetic behavior of the plasma that is able to emphasize the influence of kinetic effects on the resonance structure. Similar to *particle-in-cell*, the spectral kinetic method iteratively determines the electric field at each particle position, however, without employing any numerical grids. The optimized analytical model ensures the high efficiency of the simulation. Basically, the presented work is expected to cover the limitation of the cold plasma model, especially for the determination of the pure collisionless damping caused by kinetic effects. Notably, with the help of the spectral kinetic scheme, those energy losses can be explicitly predicted. It enables obtaining the electron temperature T_e from the half-width $\Delta\omega$ in the simulated resonance curve. That is, a formula to determine the electron temperature from the half-width is presented, which was discussed for years but unclarified. Besides, the electron density n_e can be simultaneously derived from the resonance frequency.

Furthermore, the Monte Carlo collision model is integrated into the spectral kinetic model for a realistic simulation. The simulation results are compared with the measurement in a cylindrical double inductively coupled plasma reactor, where the corresponding plasma parameters are validated by reference measurements with a Langmuir probe. Eventually, good agreements in the comparison between the kinetic simulation and the experiment demonstrate the suitability of the presented scheme. The limitations of the cold plasma model are covered by the spectral kinetic model, and the pronounced kinetic effects in the low-pressure plasmas are well explained. Consequently, the spectral kinetic model can be seen as indispensable support in the MRP-plasma system for reliable supervision and control of the plasma process.

Contents

List of Figures

List of Tables

1 Introduction

1.1 Plasma Diagnostics

Plasma science has gained increasing attention over the last decades, which is known as a multidisciplinary research area. Specifically, it is often a critical component of many disciplines, including astrophysics and space science, spectroscopy, surface and interface physics, and biomedical physics. Without doubt, plasma physics embraces almost the full breadth of the subfields in physics, and it also leads to the future innovation and development of these fields. As one of the widest utilized applied sciences, the diversity of its industrial application is remarkable, such as the well-known semiconductor manufacturing, intricate surface processing, plasma sterilization and disinfection, and realization of controlled nuclear fusion, just to name a few. In other words, plasma science has an extraordinarily broad impact in numerous industrial fields. The study of plasma technology is therefore of great interest within the general field of plasma science [1, 2].

The term plasma was first introduced by Irving Langmuir in the context of ionized gases in 1929 [3]. A plasma is a partially or completely ionized gas exhibiting collective behavior, which contains electrons, ions, and neutrals. In general, the plasma density n and the electron temperature T_e are essential for the characterization of a plasma. The plasma density usually refers to the number of charged species per unit of volume in the plasma bulk, and the electron temperature directly depends on the electron energy. Both parameters can vary significantly [4, 5]. Central to both experiment and theory is the accurate measurement of those key plasma parameters. Diagnostics are not only advantageous for the understanding of plasmas in the process, which provide better models and related simulations, but also beneficial for the process control in industrial applications. Therefore, one of the major topics to be investigated in plasma technology is the supervision and control of plasma. Figure 1.1 shows a schematic block diagram of the process control system. In order to achieve predictable and reproducible process results, many diagnostic techniques are proposed, especially due to the growing demand in industrial applications. However, only a few are applicable in technical plasmas and able to fulfill the industrial requirements. Thus, process control employing diagnostic systems still remains a challenging and vivid discipline. The diagnostic system must be robust and stable. Besides, it must minimally perturb the plasma process and be insensitive against perturbation by the process. Moreover, particularly important for industrial applications is the economical price of diagnostic systems. Furthermore, the spatially and time-resolved measurements of the controlled plasma process with a fast evaluation are required.

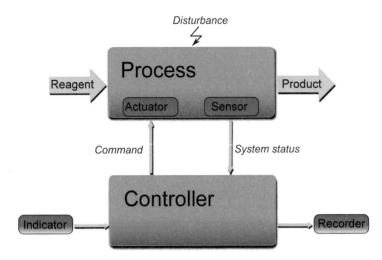

Figure 1.1: Control loop with disturbance of the process: The controller monitors the controlled process via a sensor. Depending on the captured system status, the adjustment can be executed by the actuator. The indicator sets the goal of the process, whereas the recorder collects the data.

The effort has been made by empirical optimization of plasma process to find the defined external parameters, such as voltage or current, which can be manipulated over the whole process according to the requirements. Unfortunately, controlled external parameters can cause a drift in the plasma process [6]. Thus, the internal parameters, such as the electron density or the electron temperature, need to be taken into account to closely monitor and adjust the plasma process.

1.2 Active Plasma Resonance Spectroscopy

Many different existing techniques are available for measuring the spatial profile and determination of various plasma parameters. Only few are suitable for industrial settings. One of the well-known concepts to industry-compatible plasma diagnostics is the so-called active plasma resonance spectroscopy (APRS). The method can be used in areas from moderate to very low pressures, which is often given in the industrial plasmas. The precise control is achieved by in-situ diagnostics. The concept of APRS is to utilize the natural ability of plasmas to resonate on or near the electron plasma frequency ω_{pe}, which was initially investigated back in 1929 [7]. In Figure 1.2, the idea of APRS is depicted: a signal is coupled into the plasma via a radio-frequency (RF) fed probe, and the response of the plasma is recorded in a certain frequency range [8]. Then some important plasma parameters can be determined by a mathematical model.

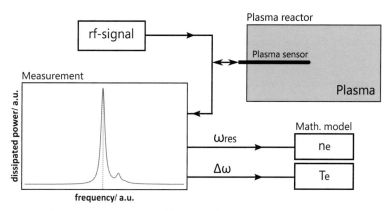

Figure 1.2: Schematical depiction of APRS: a radio-frequency signal is coupled into plasma via a probe, and the spectral response is recorded and analyzed. From the resonance curves, the corresponding plasma parameters can be determined by using a mathematical model.

To be more specific, the electron plasma frequency can be expressed in terms of the electron density,

$$\omega_{\mathrm{pe}} = \sqrt{\frac{e^2 n_{\mathrm{e}}}{\varepsilon_0 m_{\mathrm{e}}}}. \tag{1.1}$$

According to this definition, since the elementary charge e, the vacuum permittivity ε_0, and the electron mass m_{e} are constants, the electron density can be calculated if the electron plasma frequency is identified [9, 10]. Similarly, the ion plasma frequency ω_{pi} relates to the ion mass m_{i} and the ion density n_{i},

$$\omega_{\mathrm{pi}} = \sqrt{\frac{e^2 n_{\mathrm{i}}}{\varepsilon_0 m_{\mathrm{i}}}}. \tag{1.2}$$

However, the ions are much heavier than the electrons ($m_{\mathrm{i}} \gg m_{\mathrm{e}}$), which causes a much smaller ion plasma frequency ($\omega_{\mathrm{pi}} \ll \omega_{\mathrm{pi}}$). It indicates that the electron plasma frequency is of particular importance to plasma dynamics. The proportionality of the electron plasma frequency and the resonance frequency ω_{res} can be assumed, which writes

$$\omega_{\mathrm{res}}^2 \propto \omega_{\mathrm{pe}}^2 \propto n_{\mathrm{e}}. \tag{1.3}$$

Therefore, from the obtained resonance response, the resonance frequency of the plasma can be evaluated in a mathematical model. Additionally, the electron temperature T_{e} is linked to the collision frequency ν. It can also be interpreted from the measurements. Hence, the determination of these important internal plasma parameters can be directly achieved in this method.

1.3 Multipole Resonance Probe

Plasma probes are, in most cases, easy to use and give access to plasma parameters. The most renowned diagnostic probe is the Langmuir probe (LP), which was developed in the 1920s [11]. A Langmuir probe consists of a bare wire or an isolated metallic. To ensure reliable results, the tip must be conditioned so as not to interfere with the plasma nor be destroyed by it. The current can be measured at various applied voltages to obtain the current-voltage (I-V) characteristics [12]. This time-honored method is used in a wide variety of industrial or laboratory plasma devices. The theory of LP is well documented [13, 14]. In principle, plasma characteristics, such as n_e, T_e, are extracted from the I-V curves by the LP system. However, it often presents several challenges regarding the complexity of the processes [15, 16].

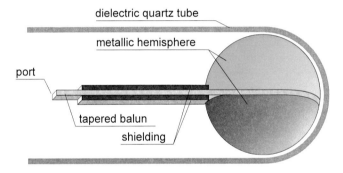

Figure 1.3: Prototype of the MRP: The probe consists of two metallic hemispheres. The total diameter is 8mm. It is symmetrically driven via a tapered balun transformer, and it can be covered with a dielectric in a cylindrical quartz tube.

For a reliable real-time measurement of the internal plasma parameters, the multipole resonance probe (MRP) was proposed in 2008 as one of the realizations of APRS [17]. The device is a radio-frequency driven probe of a particular spherical design, which is powered in a frequency range approximately between 100 MHz and 10 GHz. The prototype of the MRP is shown in Figure 1.3 [18]. It consists of a spherical probe head and a holder. The MRP's head is comprised of two conducting metallic hemispheres, which constitute the electrically symmetric electrodes. These two hemispheres are separated by a dielectric layer, and fixed to a holder that contains the RF-supply. The probe is inserted into a quartz tube in order to isolate it from plasma. The setup of the MRP provides two important features: its electrical behavior is symmetric regarding the mapping, and also its geometry is approximately symmetric.

The MRP is developed in a collaboration of four institutes at the Ruhr-University Bochum within the framework of the project PluTO and PluTO+ (Plasma and optical technologies), which is funded by the Federal Ministry of Education and Research (BMBF). Many dissertations and publications are contributed to the improvement of the MRP during the last decade, which can be seen as a great success of the project. The related work during this period is summarized in [19]. It is a long and challenging process to close the theory-practice gap. In July 2020, a start-up with the name "House of Plasma" was launched at the Ruhr-University Bochum. The project is financially supported by the Ministry of Economic Affairs, Innovation, Digitalization and Energy of the state of North Rhine-Westphalia. Many different designs of MRP are proposed to meet the requirements in the process, e.g., in Figure 1.4, the device can be flexibly mounted to prevent shadowing [20]. Besides, other design, such as the MRP with an adjustable length of the tube, provides simple solutions for plasma diagnostics. Eventually, this innovative sensor and its applications become commercially available, which can be used for monitoring and controlling low-pressure plasma processes.

Figure 1.4: Free positionable MRP: it is the most flexible design, which is suitable for monitoring and control systems. (This photo is provided by House of Plasma.)

For the monitoring and control systems with the reduced influence on the plasma, the planer multipole resonance probe (pMRP) is designed as a minimal-invasive device based on the idea of a compact planer sensor: two electrodes, which are protected by the dielectric, are given in planer structure fed by a coaxial cable. The prototype is shown in Figure 1.5 [20], and the principal functionality is discussed in [21]. The pMRP is hidden in the plasma process since it is mountable into the reactor wall. Nevertheless, it holds the same features as the MRP. The comparable resonance peaks can be observed in [22].

While the MRP is suitable for an all-embracing characterization of the plasma bulk, the pMRP provides the alternative solution without physical presence of the probe. Furthermore, in [23], the analytic model is derived based on the cold plasma model to have a better understanding of this stationary measurement system.

Figure 1.5: planer MRP: the probe is designed with a planer structure, which can be integrated into the reactor wall. (This photo is provided by House of Plasma.)

Similar to the MRP, a variety of probes for reliable and accurate measurements have been invented, such as the microwave resonator probe [24, 25, 26], curling probe [27, 28, 29], plasma absorption probe [30, 31, 32], and many other designs [33, 34, 35, 36, 37, 38]. The resonance structure is captured and analyzed in many different approaches for understanding the resonance behavior of the plasma [39, 40, 41, 42, 43, 44]. In general, not only one resonance peak in the spectrum but a number of resonance peaks in multiple resonance structures are observed. This structure is caused by the excitation of higher resonance modes. Such phenomena can be seen in experiments and simulations, which complicates the interpretation of the measurements [45, 46]. To remedy the situation, the MRP is designed with a particular setup, which enables the unique feature: the resonances at different frequencies can be measured and well sorted, where the most important and dominant resonance, also known as the dipole resonance, is clearly identified. The analytical description of the MRP is based on the cold plasma model [17]. The plasma is treated as a frequency-dependent dielectric material. The analyzed MRP is highly idealized, where the holder is neglected. In [47], the first measurements of an MRP prototype agree with parallel Langmuir probe measurements. Thus, the feasibility of the MRP concept is proved.

In addition, the numerical simulation of the MRP with complete geometry is investigated within 3D-electromagnetic field simulations using CST Microwave Studio, where the influence of the holder is included [48, 49, 50]. A very similar resonance behavior of the plasma is demonstrated in both approaches. The analytical prediction is verified by the comparable results in the simulations. The negligible difference of the dominant dipole resonance in the comparison indicates that the idealization of the MRP is acceptable. In [21], the simulated magnitude of the electric field shows that most of the interactions in the plasma appear near the tip of the probe. Therefore, the hemisphere on the opposite side of the holder can be treated as an ideal model, and the other half of the sphere is assumed symmetric, which forms the ideal multipole resonance probe (IMRP): two ideal hemisphere electrodes are covered by a dielectric. The spherically symmetric geometry of IMRP allows a highly optimized analytical model.

These theoretical and numerical results are substantially confirmed through some further simulation and experimental work in [51,52,53]. Moreover, a variety of MRP's applications are studied, such as in the dielectric deposition processes [54], and medical sterilization processes [55]. One of the challenges for the MRP system in industrial processes is the high temperature of working environments. In some cases, it may go far above 500°C. To increase the robustness, the MRP is implemented with the so-called low temperature co-fired ceramics (LTCC) as the substrate material. Thus, the improved MRP can be seen as an industry-compatible sensor with significantly increased temperature stability [56,57].

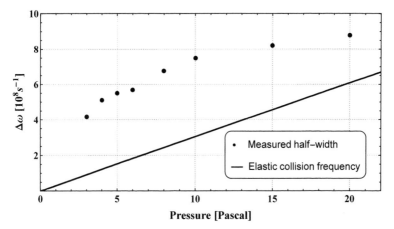

Figure 1.6: Comparison of the half-width $\Delta\omega$ between the measured spectra and the electron neutral collision frequency in the cold plasma model

Table 1.1: An overview of the related work: the difference between the idealized MRP and realistic MRP in the cold plasma model and the kinetic model is presented.

	Ideal MRP	Realistic MRP
Cold plasma model	Unrealistic physics	Unrealistic physics
	Unrealistic geometry	Realistic geometry
Related work	Analytical model	CST simulation
Kinetic model	Realistic physics	Realistic physics
	Unrealistic geometry	Realistic geometry
Related work	Spectral kinetic model	Not necessary

Although the fluid approach allows simple and effective modeling of the plasma-probe system, especially accurate for the prediction of the electron density from the simulated ω_{res}, there are apparent deviations. While the absolute position ω of the resonances was quantitatively recorded, especially for dipole mode, the half-width $\Delta\omega$ of dependent damping cannot be reproduced via the cold plasma model in low-pressure plasmas. In Figure 1.6, the comparison between the half-width of the measured spectra of the MRP and the calculated electron-neutral collision frequency based on the cold plasma model is presented. Apparently, the cold plasma model has a poor prediction in a pressure regime of a few Pa or lower. This phenomenon has been observed for over a decade. However, it has not been clarified yet. The half-width $\Delta\omega$ reflects the damping of the plasma, which is mostly contributed by the elastic electron-neutral collisions. The elastic collision frequency depends on the pressure of the background gas. In general, the collisions between the electron and neutral are dominant in the energy loss due to the low ionization degree. However, in low-pressure plasmas, the energy loss caused by the escape of the free particle becomes pronounced, which can be interpreted as kinetic effects. Since $\Delta\omega$ is crucial to determine the electron temperature, the kinetic effects can not be neglected in the low-pressure regime.

To overcome the limitation of the cold plasma model, the kinetic modeling of MRP is required. In 1966 [58], a study of the kinetic theory was presented to analyze the behavior of the conductive spherical probe, a basic understanding of energy losses from the plasma was discussed. In [59,60,61], a much more generalized kinetic model is proposed for different probes, which is valid for all pressures and arbitrary geometries. The dynamic interaction of the probe with the plasma of its influenced domain is described via functional analytic (Hilbert Space) methods, and the response function of the probe-plasma system is expressed as matrix elements of the resolvent of a defined dynamical operator. The broadening of the spectrum at low pressure is explained. Unfortunately, it remains challenging for collisionless cases, which is greatly relevant to the pure kinetic effects. Alternatively, solving the integration of the Vlasov equation can be another possible approach. However, it is cumbersome in regard to the calculations. Thus, a more straightforward particle model is necessary.

Considering the complexity of a particle-based simulation, there are demands for an optimized scheme to describe the probe-plasma system. Table 1.1 shows the overview of the study related to MRP. For the realistic model, the presence of the holder breaks the exact spherical symmetry, which leads to sophisticated mathematical models in the kinetic approach. Additionally, the plasma systems are often extremely large concerning the number of particles. Therefore, the kinetic model of the realistic MRP can be computationally costly. However, according to the aforementioned study of MRP based on the cold plasma model [47, 48], IMRP is suitable for the theoretical investigation, which simplifies the calculations for solving the kinetic model. Thus, a kinetic model of the IMRP with high computational efficiency is expected to cover the deviation of the cold plasma model in the low-pressure regime. Eventually, a much more accurate prediction of the electron temperature, in addition to the electron density, can be provided by the spectral kinetic model [8, 18], which can significantly improve the usability and reliability of the MRP in the supervision and control of the plasma process.

1.4 Scope of the Work

The intention of this work is to derive a theoretical model to describe the interactions between the particles and the influence of the MRP. In the earlier research, the numerical analysis is based on the actual design of the MRP. However, to be more suitable for the theoretical investigation, the MRP is idealized in the study of the cold plasma model. Importantly, based on the idealized MRP, the noticeable features of the MRP can also be described in the proposed spectral kinetic model. It simulates the scenario that an impulse signal is provided via the IMRP, then the response of the plasma around the probe is captured. The term spectral refers to the recorded frequency-dependent spectrum. The calculations in the frequency domain are time-consuming regarding the convergence of a sequence of periodic functions. Therefore, in the proposed scheme, the output of the simulation, such as the charge on the electrodes, is calculated in the time domain. By applying a Fourier transform, the impulse response can be expressed in the frequency domain for an efficient evaluation.

The aim of the simulation is to reveal the kinetic effects, which are absent in the cold plasma model. In reality, the electrons are deflected by the electric field due to the IMRP. The escape of the particles carries the energy out of the influenced domain, which can not be described in the fluid description. Such phenomena is interpreted as kinetic collision frequency. By the investigation of the energy losses in the plasma-IMRP model, the kinetic effects are to be explained, and the comparison between the kinetic model and the cold plasma model are to be discussed. The input parameters, taken from the theoretical predictions or the experimental observations, are used in the simulations to provide the relevant output. Then the simulation results are validated from the comparison of the corresponding measurements. Eventually, according to the input-output relation of the spectral kinetic simulation, the desired plasma parameters, such as the electron density n_e, electron temperature T_e, can be directly determined from the measured resonance curves, where the resonance frequency ω_{res}, half-width $\Delta\omega$, collision frequency ν are obtained. Furthermore, a formula to calculate the electron temperature from the half-width is proposed, which improves the accuracy and reliability of the measurements from MRP.

In particular, the following questions are to be answered in this work:

- What is the difference between the cold plasma model and the kinetic model?

- What are "kinetic effects", and why they are missing in the cold plasma model?

- What is the difference between the spectral kinetic method and the particle-in-cell method?

- Is it possible to observe kinetic effects in the simulation of the spectral kinetic model?

- What is the influence of kinetic effects to the resonance structure?

- What is the relation between the half-width $\Delta\omega$ of the resonance curves and the electron temperature T_{e}?

- Is it possible to validate the simulation against experimental data?

This dissertation is organized as follows:

- Chapter 2 gives a general description of the model: The difference between the fluid model and the kinetic model is discussed. The details of fluid description of plasma are presented, where the cold plasma model is derived. Then the unique feature of MRP is demonstrated in the simulated resonance structure from the previous study of cold plasma model. However, the limitation of the cold plasma model leads to the necessity of a kinetic investigation of the plasma-probe system.

- Chapter 3 introduces some numerical methods and the concept of computer simulation for efficiently solving the complicated particle model. Specifically, the structure of the kinetic scheme is introduced, which is similar to the well-known particle-in-cell method, however, without employing any numerical grids. The analysis and comparison of these two approaches are presented. Besides, Monte Carlo collision method is coupled with the proposed kinetic model to describe the collisions in the simulation.

- In chapter 4, the multipole resonance probe is idealized for the theoretical investigation. Then the spectral kinetic model is introduced analytically in detail: An explicit Green's function is determined as the solution of the Poisson's equation, which is expanded in the spherical harmonics. Due to the symmetry of the IMRP, a certain truncation can be executed for the simplification of the calculation.

- In chapter 5, the implementation of the simulation is presented. Regarding the initial condition, the state before the onset of the electric signal is numerically evaluated from the Boltzmann-Poisson equation to enhance the efficiency of the simulation. Besides, the setup of the simulation, such as the boundary condition, and the energy distribution of the particles are determined. By applying a Fourier transform, the simulation results in the time domain are converted into the frequency domain for further analysis.

- In chapter 6, the results of the spectral kinetic simulations are discussed in detail. Firstly, the prediction of resonance frequencies from the given electron densities is presented, which is compared with the results in the cold plasma model. Moreover, the half-widths of the simulated resonance curves are determined from the given electron temperatures, where the kinetic effects are demonstrated. Remarkably, the direct prediction of the electron temperature from the half-widths is proposed as a formula. Furthermore, the setup of the experiment is described. Eventually, these simulation results are validated by the comparison of the experimental data.

- In chapter 7, this work is concluded, and the possible future research directions are discussed.

This dissertation is partially based on the following peer-reviewed journal articles. Chapter 4 is reproduced from [8] with the permission of AIP Publishing, and parts of the results in Chapter 6 have been published in [18] under a Creative Commons Attribution (CC BY) license.

- [8] J. Gong, M. Friedrichs, J. Oberrath and R. P. Brinkmann, "Kinetic simulation of the ideal multipole resonance probe", *Journal of applied physics*, 132(6): 064502, 2022.

- [18] J. Gong, M. Friedrichs, J. Oberrath and R. P. Brinkmann, "The multipole resonance probe: Simultaneous determination of electron density and electron temperature using spectral kinetic Simulation", *Plasma Sources Science and Technology*, 31(11): 115009, 2022.

- [19] J. Oberrath, M. Friedrichs, J. Gong et al., "On the Multipole Resonance Probe: Current Status of Research and Development", *IEEE Transactions on Plasma Science*, 49(11): 3293–3298, 2021.

2 Description of the Model

Experimental physics plays a vital role in understanding natural phenomena. The measurements and observations contribute to new theories and discoveries. Besides, the experiments are inevitable for the validation of the developed numerical models, and some input parameters of the simulation are required from the measurements. However, several obstacles can not be avoided at a certain point in the experimental work. For example, the phenomena are too fast to be captured, the experimental techniques can only be carried out at limited locations in the processes, the setup is too difficult to be realized due to its extremely large or small size, or the materials under study are simply very expensive, just to name a few. However, some of these experimental obstacles can be overcome by using theoretical descriptions and numerical modeling.

Different kinds of mathematical models exist in the literature to predict plasma behavior under specific operating conditions and describe the various plasma processes. There are two most widely chosen approaches to plasma simulation, which are kinetic descriptions and fluid descriptions. The most fundamental way to describe a plasma is the kinetic model, which is also known as the particle model. As is suggested by its name, the plasma is treated as an ensemble of free particles in a microscopic description. Contrarily, a fluid model describes the plasma in terms of average, macroscopic quantities, such as density, momentum, and energy. Basically, in kinetic simulation, the kinetic behavior of the plasma is computed, which contains the individual particle motions, whereas the fluid simulation models the coarse-grained aggregate behavior.

Each of these mathematical models is characterized by its specific advantages and disadvantages. Kinetic models are able to calculate the entire dynamic behavior of plasma accurately with a very high degree of reliability. However, a common drawback is the longer calculation time and higher computational cost. Conversely, fluid models are relatively fast techniques to calculate the electric field in a self-consistent manner. But the fluid description reaches the limitation that plasma cannot be considered a continuum at low pressure. To be more specific, the loss of energy caused by electron-neutral collisions is less dominant than the energy gain due to the electric or magnetic field. It indicates that the plasma is therefore not in thermal equilibrium, which makes the fluid model less suitable for low-pressure plasmas. Therefore, depending on the investigated plasma, the type of the applied model is decided. Even in some cases, the hybrid models, which are the combinations of several types of models, are used to provide accurate results in an efficient way. In this approach, the advantages of different models are utilized while their limitations are compensated.

In this chapter, a brief discussion of the most frequently used models for the description of the plasma behavior is presented. The mathematical description of the technical plasmas is derived, including the consideration of the different types of collisions. Moreover, the previous study of MRP based on the cold plasma model is introduced. However, the results with obvious drawbacks are demonstrated, which are to be analyzed in the following.

2.1 Kinetic Description of Plasma

In the kinetic theory, the plasma is commonly understood as an ensemble of different individual particles, which are described on the basis of classical mechanics and electrodynamics. To more specific, based on this concept, the plasma can be treated as a collection of charged particles characterized by mass m and charge q. In general, a distribution function $f(\boldsymbol{r}, \boldsymbol{v}, t)$ in the $(6+1)$-dimensional phase space $(\boldsymbol{r}, \boldsymbol{v}, t)$ is introduced. We assume a total number of N particles are in a six-dimensional volume of phase space, which includes components v_1, v_2, v_3 in the velocity coordinates and x_1, x_2, x_3 in spatial coordinates. As functions of time t, velocity $\boldsymbol{v}_k(t)$, and position vector $\boldsymbol{r}_k(t)$ are used to define particle k. Basically, all particles of the same species are summarized by the assigned index α. Here, the superscript index m refers to microscopic. Then, the microscopic distribution function $f_\alpha^{\mathrm{m}}(\boldsymbol{r}, \boldsymbol{v}, t)$ can be written in the form of the δ-function on this phase space, which describes the location and velocity of each particle [62]

$$f_\alpha^{\mathrm{m}}(\boldsymbol{r}, \boldsymbol{v}, t) = \sum_{k=1}^{N_\alpha} \delta^{(3)}(\boldsymbol{r} - \boldsymbol{r}_k(t)\delta^{(3)}(\boldsymbol{v} - \boldsymbol{v}_k(t)), \tag{2.1}$$

Microscopically, the fields include the external electric and magnetic fields and the self-consistent fields from the particles. These fields within the plasma are described by Maxwell's equations:

$$\nabla \cdot \boldsymbol{E}^{\mathrm{m}}(\boldsymbol{r}, t) = \frac{\rho^{\mathrm{m}}(\boldsymbol{r}, t)}{\varepsilon_0}, \tag{2.2}$$

$$\nabla \times \boldsymbol{E}^{\mathrm{m}}(\boldsymbol{r}, t) = -\frac{\partial \boldsymbol{B}^{\mathrm{m}}(\boldsymbol{r}, t)}{\partial t}, \tag{2.3}$$

$$\nabla \cdot \boldsymbol{B}^{\mathrm{m}}(\boldsymbol{r}, t) = 0, \tag{2.4}$$

$$\nabla \times \boldsymbol{B}^{\mathrm{m}}(\boldsymbol{r}, t) = \mu_0 \boldsymbol{j}^{\mathrm{m}}(\boldsymbol{r}, t) + \mu_0 \varepsilon_0 \frac{\partial \boldsymbol{E}^{\mathrm{m}}(\boldsymbol{r}, t)}{\partial t}. \tag{2.5}$$

To cover all the particles, it is necessary to integrate over the entire phase space. Then, the microscopic charge density is given as

$$\rho^{\mathrm{m}}(\boldsymbol{r}, t) = \sum_{\alpha} \int_{V} q_{\alpha} f_{\alpha}^{\mathrm{m}}(\boldsymbol{r}, \boldsymbol{v}, t) \mathrm{d}V, \tag{2.6}$$

and the microscopic current density is

$$\boldsymbol{j}^{\mathrm{m}}(\boldsymbol{r}, t) = \sum_{\alpha} \int_{V} q_{\alpha} \boldsymbol{v} f_{\alpha}^{\mathrm{m}}(\boldsymbol{r}, \boldsymbol{v}, t) \mathrm{d}V. \tag{2.7}$$

Taking the time derivative of the distribution function 2.1, the time evolution of the plasma can be expressed by the Klimontovich equation [63], which is in terms of conservation law

$$\frac{\partial f_{\alpha}^{\mathrm{m}}(\boldsymbol{r}, \boldsymbol{v}, t)}{\partial t} + \boldsymbol{v} \cdot \nabla f_{\alpha}^{\mathrm{m}}(\boldsymbol{r}, \boldsymbol{v}, t) + \frac{q_{\alpha}}{m_{\alpha}} (\boldsymbol{E}^{\mathrm{m}} + \boldsymbol{v} \times \boldsymbol{B}^{\mathrm{m}}) \cdot \nabla_{\mathrm{v}} f_{\alpha}^{\mathrm{m}}(\boldsymbol{r}, \boldsymbol{v}, t) = 0. \tag{2.8}$$

This microscopic description indicates the incompressibility of the plasma. In principle, if the initial condition of $\boldsymbol{v}_k(t)$ and $\boldsymbol{r}_k(t)$ at $t = 0$ is given, the exact trajectories of all particles in the phase space can be calculated. However, being equivalent to solving all the particle motion, the Klimontovich equations are practically not solvable. In other words, the difficulty of solving such complex equations prevents us from obtaining anything significant. The situation can be remedied by averaging the equation over a small volume, where discrete particles in the phase space are smoothed. However, the features of plasma behavior are still resolved. Then, the averaged Klimontovich equation, also known as the kinetic equation, can be established

$$\frac{\partial f_{\alpha}(\boldsymbol{r}, \boldsymbol{v}, t)}{\partial t} + \boldsymbol{v} \cdot \nabla f_{\alpha}(\boldsymbol{r}, \boldsymbol{v}, t) + \frac{q_{\alpha}}{m_{\alpha}} (\boldsymbol{E} + \boldsymbol{v} \times \boldsymbol{B}) \cdot \nabla_{\mathrm{v}} f_{\alpha}(\boldsymbol{r}, \boldsymbol{v}, t) = \langle f_{\alpha} \rangle_{\mathrm{c}}. \tag{2.9}$$

The macroscopic fields can be calculated by coupling with corresponding Maxwell equations

$$\varepsilon_0 \nabla \cdot \boldsymbol{E}(\boldsymbol{r}, t) = \sum_{\alpha} q_{\alpha} \int_{V} f_{\alpha}^{\mathrm{m}}(\boldsymbol{r}, \boldsymbol{v}, t) \mathrm{d}V, \tag{2.10}$$

$$\nabla \times \boldsymbol{E}(\boldsymbol{r}, t) = -\frac{\partial \boldsymbol{B}(\boldsymbol{r}, t)}{\partial t}, \tag{2.11}$$

$$\nabla \cdot \boldsymbol{B}(\boldsymbol{r}, t) = 0, \tag{2.12}$$

$$\frac{1}{\mu_0} \nabla \times \boldsymbol{B}(\boldsymbol{r}, t) = \sum_{\alpha} q_{\alpha} \int_{V} \boldsymbol{v} f_{\alpha}^{\mathrm{m}}(\boldsymbol{r}, \boldsymbol{v}, t) \mathrm{d}V + \varepsilon_0 \frac{\partial \boldsymbol{E}(\boldsymbol{r}, t)}{\partial t}. \tag{2.13}$$

Therefore, a macroscopic description is realized by the continuous distribution function $f_\alpha(r, v, t)$ in $6 + 1$-dimensional phase space, which characterizes each particle species α in the plasma. The product $f_\alpha drdv$ refers to the average number of particles. The position and velocity of those particles are in the volume $drdv$ centered at (r, v). In the kinetic equation (2.9), the average behavior is on the left-hand side, where the collective effects of the plasma are described, whereas the particle-like behavior, which is due to collisions and correlations, is on the right-hand side. Specifically, the second term corresponds to the flow of particles, and the third term is the field term. Besides, regarding the collision term $\langle f_\alpha \rangle_c$, it does not initially indicate any specific properties to a plasma. In fact, there are various approaches to describe this in the literature [64, 65]. That is to say, the kinetic equation forms the basis of the kinetic plasma theory. However, it only gets its final form depending on the choice of the collision term, e.g., derived from the general kinetic equation; by using a special integral approach that considers the collisions between electrons and neutral gas particles, the Boltzmann equation can be obtained, which plays a significant role, especially for the consideration of the kinetic description of the low-temperature plasma [66]. As a nonlinear partial integro-differential equation, the solution of it is usually not analytically possible. Hence, some suitable approximations are required. Nevertheless, it can be cumbersome in regard to the calculations for the complicated MRP-plasma system. Alternatively, a particle-based model with the numerical approach becomes a reliable tool due to its scalability with the development of computer science. The specific numerical scheme in this work is introduced in Chapter 4.

Apparently, the kinetic approach of plasma is spectacularly inefficient in terms of calculations. As is introduced, the distribution function contains vastly too much information about all the particles to obtain the constitutive relations. The complexity of the kinetic equation makes the model analytically unsolvable without any further approximation. Therefore, the fluid approach of plasma provides the possibility to impose some efficiency in solving the plasma model.

In the fluid description, the macroscopic quantities are used to characterize the plasma. These local parameters, such as the particle density, and flow velocity, can be captured from the microscopic behavior of the particles. Hence, the fluid equations possess the simplicity in computer simulations compared to the kinetic description. Besides, fluid variables are comparatively easy to obtain in the measurements. Therefore, the mentioned intrinsic advantages indicate the necessity investigating of the fluid description.

2.2 Fluid Description of Plasma

2.2.1 Moments of the Distribution Function

By using the Boltzmann equation (2.9), the macroscopic behavior of plasma is captured from the microscopic behavior of its particles [67]. From the distribution function f_α, some macroscopic physical parameters, like particle density or particle flux density, can be obtained by the integral over the velocity space as so-called moments. The i-th moment of the ensemble-averaged distribution function is in the form of an integral, which is written as

$$M_\alpha^{(i)}(f_\alpha) = \int \underbrace{\boldsymbol{v} \cdots \cdot \boldsymbol{v}}_{i} f_\alpha \mathrm{d}^3 v, \qquad (2.14)$$

with i factors of $\boldsymbol{v}$. Here, it is given that $i \in \{0, 1, 2, \cdots, \infty\}$, $\boldsymbol{v}^i$ (i-fold) denotes an i-fold tensor M_α^i [68]. That is, some important features of a plasma described by a distribution function are given by those moments. It is obvious that considering all the moments leads to an infinitely complicated system. Therefore, the description of plasma is usually restricted to only several moments.

First, for $i = 0$, the particle density $n_\alpha = n_\alpha(\boldsymbol{r}, t)$ can be obtained

$$M_\alpha^{(0)} = n_\alpha = \int f_\alpha \mathrm{d}^3 v. \qquad (2.15)$$

Then, for $i = 1$, the particle flux density $\boldsymbol{\Gamma}_\alpha = \boldsymbol{\Gamma}_\alpha(\boldsymbol{r}, t)$ is written as

$$M_\alpha^{(1)} = \boldsymbol{\Gamma}_\alpha = n_\alpha \boldsymbol{u}_\alpha = \int \boldsymbol{v} f_\alpha \mathrm{d}^3 v. \qquad (2.16)$$

Here, the quantity $\boldsymbol{u}_\alpha = \boldsymbol{u}_\alpha(\boldsymbol{r}, t)$ is the flow velocity. Similarly, for $i = 2$, the second-order moment, multiplied by the mass m_α, gives the stress tensor $\underline{\underline{P}}_\alpha = \underline{\underline{P}}_\alpha(\boldsymbol{r}, t)$, which describes the flow of momentum

$$M_\alpha^{(2)} = \underline{\underline{P}}_\alpha = m_\alpha \int \boldsymbol{v}\boldsymbol{v} f_\alpha \mathrm{d}^3 v. \qquad (2.17)$$

This quantity can be measured in the rest frame of the species under consideration, which is the so-called pressure tensor. It is important to distinguish between the particle velocity $\boldsymbol{v}$ and the flow velocity $\boldsymbol{u}_\alpha$. The random velocity of particles, also known as the relative velocity, is defined as

$$\boldsymbol{w}_\alpha = \boldsymbol{v} - \boldsymbol{u}_\alpha, \qquad (2.18)$$

accordingly, the pressure tensor $\underline{\underline{p}}_\alpha = \underline{\underline{p}}_\alpha(\boldsymbol{r}, t)$ is denoted by

$$\underline{\underline{p}}_\alpha = m_\alpha \int \boldsymbol{w}_\alpha \boldsymbol{w}_\alpha f_\alpha \mathrm{d}^3 v. \qquad (2.19)$$

The relation between the stress tensor $\underline{\underline{P_\alpha}}$ and the pressure tensor $\underline{\underline{p_\alpha}}$ can be easily derived as

$$\underline{\underline{P_\alpha}} = m_\alpha n_\alpha \boldsymbol{u}_\alpha \boldsymbol{u}_\alpha + \underline{\underline{p_\alpha}}. \tag{2.20}$$

The elements of the pressure tensor $\underline{\underline{p_\alpha}}$ can be represented in a 3×3-matrix as follows:

$$\underline{\underline{p_\alpha}} = \begin{pmatrix} p_{xx} & p_{xy} & p_{xz} \\ p_{yx} & p_{yy} & p_{yz} \\ p_{zx} & p_{xy} & p_{zz} \end{pmatrix} \approx \begin{pmatrix} p_{xx} & 0 & 0 \\ 0 & p_{yy} & 0 \\ 0 & 0 & p_{zz} \end{pmatrix}. \tag{2.21}$$

At this point, it makes sense to neglect certain elements due to symmetry properties in thermodynamic equilibrium, which are $p_{xy} = p_{yx}$, $p_{xz} = p_{zx}$, $p_{yz} = p_{zy}$. However, the diagonal elements are obviously nonzero, which must be considered individually. Therefore, the trace of the pressure tensor $\underline{\underline{p_\alpha}}$ measures the scalar pressure p_α

$$p_\alpha = \frac{1}{3}\operatorname{tr}\left(\underline{\underline{p_\alpha}}\right) = \frac{1}{3}\left(p_{xx} + p_{yy} + p_{zz}\right). \tag{2.22}$$

Furthermore, $i = 2$ gives the third-order moment measuring the energy flux density $\underline{\underline{\underline{Q_\alpha}}} = \underline{\underline{\underline{Q_\alpha}}}(\boldsymbol{r}, t)$, which writes

$$M_\alpha^{(2)} = \underline{\underline{\underline{Q_\alpha}}} = \frac{1}{2} m_\alpha \int \boldsymbol{v}\boldsymbol{v}\boldsymbol{v} f_\alpha \mathrm{d}^3 v. \tag{2.23}$$

If the energy flux density $\underline{\underline{\underline{Q_\alpha}}}$ is measured in the rest frame, heat flux density $\underline{\underline{\underline{q_\alpha}}} = \underline{\underline{\underline{q_\alpha}}}(\boldsymbol{r}, t)$ is obtained

$$\underline{\underline{\underline{q_\alpha}}} = \frac{1}{2} m_\alpha \int \boldsymbol{w}_\alpha \boldsymbol{w}_\alpha \boldsymbol{w}_\alpha f_\alpha \mathrm{d}^3 v. \tag{2.24}$$

The second-order and third-order moments measured in different frames are related. By substitution, the energy flux density can be expressed as

$$\underline{\underline{\underline{Q_\alpha}}} = \underline{\underline{P_\alpha}}\boldsymbol{u}_\alpha + \underline{\underline{p_\alpha}}\boldsymbol{u}_\alpha + \underline{\underline{\underline{q_\alpha}}}, \tag{2.25}$$

which can now also be brought to the vector form

$$\boldsymbol{Q}_\alpha = \operatorname{tr}\left(\underline{\underline{P_\alpha}}\right)\boldsymbol{u}_\alpha + \operatorname{tr}\left(\underline{\underline{p_\alpha}}\right)\boldsymbol{u}_\alpha + \boldsymbol{q}_\alpha. \tag{2.26}$$

Normally it is sufficient to formulate the moments of the distribution function up to the third order since higher orders for plasma physics of low-temperature are challenging to interpret physically.

2.2.2 Fluid Dynamics

While in the kinetic theory, as described in Section 2.1, single-particle dynamics is chosen as the starting point for investigations. Fluid dynamics describes a plasma as a continuous medium on a macroscopic level. That is, a simplification of the kinetic theory consists in describing the plasma in terms of its macroscopic properties. Instead of the individual particle dynamics, the collective behavior becomes prominent. The interaction between different particle types in the overall system is considered. Since the different species in this assumption behave like liquids penetrating each other, this model is called the fluid model. The transition from the kinetic equation to the fluid description is achieved by forming the moments in the velocity space. In the simplest case, this approach provides a model that contains only an electron, a positive ion, and a neutral gas fluid. In such hydrodynamic models, the investigated plasma is characterized by ensemble-averaged variables, such as density, average velocity, temperature, or pressure. In general, these quantities come directly from the distribution function. In principle, the complete information about the distribution $f_\alpha(\boldsymbol{r}, \boldsymbol{v}, t)$ can be derived from the totality of all equations. However, typically it is sufficient to consider only several leading moments.

The discussion of fluid dynamics starts from a moment expansion of the Boltzmann equation (2.9)

$$\frac{\partial f_\alpha(\boldsymbol{r}, \boldsymbol{v}, t)}{\partial t} + \boldsymbol{v} \cdot \nabla f_\alpha(\boldsymbol{r}, \boldsymbol{v}, t) + \frac{q_\alpha}{m_\alpha} (\boldsymbol{E} + \boldsymbol{v} \times \boldsymbol{B}) \cdot \nabla_{\mathrm{v}} f_\alpha(\boldsymbol{r}, \boldsymbol{v}, t) = \langle f_\alpha \rangle_{\mathrm{c}}. \qquad (2.27)$$

Taking the moment for $k = 0$, the equation for the particle density is determined by integration over the entire velocity space. The first term gives

$$\int \frac{\partial f_\alpha(\boldsymbol{r}, \boldsymbol{v}, t)}{\partial t} \mathrm{d}^3 v = \frac{\partial n_\alpha(\boldsymbol{r}, \boldsymbol{v}, t)}{\partial t}. \qquad (2.28)$$

Then, the second term is

$$\int \boldsymbol{v} \cdot \nabla f_\alpha(\boldsymbol{r}, \boldsymbol{v}, t) \, \mathrm{d}^3 v = \nabla \cdot \int \boldsymbol{v} f_\alpha(\boldsymbol{r}, \boldsymbol{v}, t) \mathrm{d}^3 v = \nabla \cdot (\boldsymbol{u}_\alpha n_\alpha). \qquad (2.29)$$

Regarding the third term, by using Gauss' theorem, the expression can be written in the form of a surface integral over the velocity space [62]. To remain the total energy finite, the vanishing of fields is assumed at infinity where no particles reside. The integrals are therefore identically zero.

As a result, the continuity equation assumes the following form:

$$\frac{\partial n_\alpha}{\partial t} + \nabla \cdot (\boldsymbol{u}_\alpha n_\alpha) = S_\alpha \qquad (2.30)$$

or

$$\frac{\partial n_\alpha}{\partial t} + \nabla \cdot \boldsymbol{\Gamma}_\alpha = S_\alpha \tag{2.31}$$

with

$$S_\alpha = \int \langle f_\alpha \rangle_c \mathrm{d}^3 v. \tag{2.32}$$

Here, S_α denotes the net change of the particle density, which involves the gain and loss for this quantity.

It is not sufficient to determine the particle density n_α with only the particle conservation since the flow velocity $\boldsymbol{u}_\alpha$ appears as the undermined variable. In order to calculate $\boldsymbol{u}_\alpha$, the higher-order moments must be formed. To be more specific, the kinetic equation is multiplied by $m_\alpha \boldsymbol{v}_\alpha$ and then integrated over the velocity space. The resulting expression is known as the momentum conservation equation, which writes

$$\frac{\partial}{\partial t}(m_\alpha n_\alpha \boldsymbol{u}_\alpha) + \nabla \cdot \left(m_\alpha n_\alpha \boldsymbol{u}_\alpha \boldsymbol{u}_\alpha + \underline{\underline{p_\alpha}} \right) = n_\alpha q_\alpha \left(\boldsymbol{E} + \boldsymbol{u}_\alpha \times \boldsymbol{B} \right) + \boldsymbol{\pi}_{\alpha,c} \tag{2.33}$$

or

$$m_\alpha \frac{\partial \boldsymbol{\Gamma}_\alpha}{\partial t} + \nabla \cdot \underline{\underline{P_\alpha}} = n_\alpha q_\alpha \left(\boldsymbol{E} + \boldsymbol{u}_\alpha \times \boldsymbol{B} \right) + \boldsymbol{\pi}_{\alpha,c}. \tag{2.34}$$

The momentum conservation equation describes the change in the momentum over time, which is essentially caused by several different aspects. The physical interpretation of the first term is the particle gains or losses. Additionally, the pressure tensor $\underline{\underline{p_\alpha}}$ relates to the particle energies or shear forces. On the right-hand side, the force is caused by the external fields. Besides, the changes due to collisions are represented by quantities $\boldsymbol{\pi}_{\alpha,c}$.

However, a mathematically challenging problem arises due to the hierarchical structure of the equation system. In order to solve equation (2.33), the pressure tensor $\underline{\underline{p_\alpha}}$, a second-order moment, must be identified, which can only be determined from the balance equation. Apparently, each balance equation represents an algebraically underdetermined differential equation, which requires the next higher moment. Therefore, the expansion must end with a certain order depending on the problem under investigation. Eventually, a self-consistent closed system of differential equations is expected from the appropriate assumptions. For example, in the following, the cold plasma model is to be derived according to the discussed equations, which are very often used in the modeling of technical plasma, including the previous study of MRP [17, 21, 22, 47, 48, 49, 50, 51]. Alternatively, particle-based simulation methods can calculate the distribution functions self-consistently and correctly reflect the dynamics of the charged particles. With the rapid development of computer science, particle-based simulation has been receiving increasing attention. A brief introduction of numerical simulation methods is presented in Section 3.

2.2.3 Cold Plasma Model

On the basis of the afore presented fluid description of plasma, a model for cold plasma is developed in this section. The starting point of the derivation is the balance equations, which result from the kinetic equation with the consideration of zeroth- and first-order moments ($k = 0$ and $k = 1$). Therefore, the cold plasma model is within the frame of the continuity equation (2.30) and the momentum conservation equation (2.33)

$$\frac{\partial n_\alpha}{\partial t} + \nabla \cdot (\boldsymbol{u}_\alpha n_\alpha) = S_\alpha, \tag{2.35}$$

$$\frac{\partial}{\partial t}(m_\alpha n_\alpha \boldsymbol{u}_\alpha) + \nabla \cdot \left(m_\alpha n_\alpha \boldsymbol{u}_\alpha \boldsymbol{u}_\alpha + \underline{\underline{p_\alpha}}\right) = n_\alpha q_\alpha \left(\boldsymbol{E} + \boldsymbol{u}_\alpha \times \boldsymbol{B}\right) + \boldsymbol{\pi}_{\alpha,\mathrm{c}}. \tag{2.36}$$

In accordance with these presented equations, the hierarchical structure of the system requires the pressure tensor $\underline{\underline{p_\alpha}}$ to obtain the solution, which indicates that certain assumptions are necessary for the balance equations.

In general, the cold plasma model is suitable for describing the particle dynamics of technical plasmas, which are often operated with high-frequency signals. For example, the radio frequency is utilized in the case of MRP. Notably, under this condition, the dynamics relatively depend on the type of the considered particle. It is defined that the index α gives the type of the particle. As is introduced in Section 1.2, ions are so much heavier than electrons. The ion mass typically goes up to 10^4 times larger, which means ions and electrons interact very differently with the external electric and magnetic fields. With the distinct contrast $\omega_{\mathrm{pi}} \ll \omega_{\mathrm{pe}}$, the light and fast electrons can follow the changes in the fields and gain net energy, whereas the ions are not strongly influenced due to the slow reaction. In a nutshell, the power coupling into the plasma is first detected by the electron and then ions; however only electrons actively respond to it. Thus, focusing on the independent dynamics of electrons in the plasma is adequate. Therefore, the balance equations of electrons are of particular interest. That is, $\alpha = \mathrm{e}$ is considered in the following.

The term cold refers to the fact that the thermal velocity of the particles in the plasma is small compared to the speed of light,

$$v_{\mathrm{th},\alpha} \ll c. \tag{2.37}$$

Taking this factor into account, a vanishing particle temperature $T_\alpha \approx 0$ can be assumed. If the plasma is at zero temperature, according to Amontons' Law, the pressure term in the equations for fluid motion will also be zero [69]. Specifically, the ratio between the speed of light to the thermal velocity of electrons is approximately $c/v_{\mathrm{th,e}} \approx 300$. Thus, the slow phenomena can be neglected. In the formalism of the cold plasma model, there is no transport of particles due to their thermal energy. For electrons, in the momentum conservation equation, the phenomena described by $\underline{p_e}$, such as pressure changes and plasma sound waves, occur at relatively slow speeds and therefore are negligible. Eventually, a closed system of equations is obtained with the vanishing of the pressure tensor. In addition, the nonlinear term $\nabla \cdot (m_e n_e \boldsymbol{u_e} \boldsymbol{u_e})$ is not taken into account since the electron mass is rather small. Besides, on the right-hand side of the equation, external forces are caused by the electromagnetic fields $\boldsymbol{E}$ and $\boldsymbol{B}$. In order to obtain a consistent system, the plasma model must be coupled to adequate field equations, such as the Maxwell equations. As a simplified form, it is convenient to adopt the electrostatic approximation, which is introduced in Section 2.2.4. The momentum conservation equation for electrons is written

$$\frac{\partial}{\partial t}(m_e n_e \boldsymbol{u_e}) = n_e q_e \boldsymbol{E} + \boldsymbol{\pi}_{e,c}. \tag{2.38}$$

Besides, processes that affect the change in particle density, such as ionization, recombination, and other chemical collision processes, can be neglected due to the significantly longer time scales than the considered dynamics, which indicates that S_e can be set as zero. Then the continuity equation follows

$$\frac{\partial n_e}{\partial t} + \nabla \cdot (\boldsymbol{u_e} n_e) = 0. \tag{2.39}$$

It is possible to characterize the collision behavior of the electrons by a collision frequency ν. That is, the change in momentum of the electrons occurs through collisions with the effective collision frequency, which depends on the process pressure, the electron temperature, and the neutral gas temperature. The related expression is given as

$$\frac{e}{m_e}\boldsymbol{\pi}_{\alpha,c} = -n_e \nu \boldsymbol{u_e} e = \nu \boldsymbol{j} \tag{2.40}$$

Essentially, there are two different types of collisions, which will be discussed in detail in Section 2.2.5. Here, the current density of electrons is given as

$$\boldsymbol{j} = -e n_e \boldsymbol{u_e}. \tag{2.41}$$

The charge density can be expressed in terms of electron and ion density

$$\rho = e\left(n_i - n_e\right), \tag{2.42}$$

where the slow reaction of the ions indicates

$$\frac{\partial n_i}{\partial t} = 0. \tag{2.43}$$

Thus, the time dependence of the ion density can be neglected, and equation (2.39) yields at

$$\frac{\partial \rho}{\partial t} = -\nabla \cdot \boldsymbol{j}. \tag{2.44}$$

Inserting equation (2.40) into equation (2.38), as a special form of conservation of momentum, the radio-frequency form of generalized Ohm's law can be obtained

$$\frac{\partial \boldsymbol{j}}{\partial t} = \varepsilon_0 \omega_{\mathrm{pe}}^2 \boldsymbol{E} - \nu \boldsymbol{j} \tag{2.45}$$

with the expression of electron plasma frequency $\omega_{\mathrm{pe}} = \sqrt{e^2 n_{\mathrm{e}}/\varepsilon_0 m_{\mathrm{e}}}$.

Consequently, the dynamics of the electrons can be described in the cold plasma approximation, which takes into account the electrons' inertia and the momentum loss caused by elastic collisions with neutrals. The underlying simplifications may differ depending on the specific problems. However, many studies examining the high-frequency behavior of technical plasmas have shown that these approximations are absolutely justified [70, 71]. In fluid dynamics, the continuity equation states that the accumulation (or loss) of particle numbers equals the difference between inflow and outflow. It is essentially a statement of the law of conservation of the number of particles. Additionally, the conservation of momentum states that the change in the electron current over time is caused by an acceleration of the electrons in the applied electric field and the impact of the collisions with the background gas, which can be interpreted as friction. In summary, the derived model corresponds to the idea of the Drude theory, which is conceptually analogous to the description of charge transport in solids.

2.2.4 Electrostatic Approximation

As stated previously in Section 2.1, the full set of Maxwell equations reveals how electric and magnetic fields are generated by charges, currents, and changes in the fields over time. It consists of four equations that determine the fields, which are Gauss's law (2.2), Faraday's law of induction (2.3), Gauss's law for magnetism (2.4), and Ampère's circuital law (2.5). However, it is not always necessary to take the full Maxwell equations into account. The so-called electrostatic approximation can be adopted in the condition that the magnetic field is constant or at least slowly varying. That is, the changes in the magnetic field $\boldsymbol{B}$ over time are negligible. Notably, the length scale is smaller than the skin depth. Therefore no skin effects are considered. The assumption of the vanishing of induction term in Faraday's law of induction is justified since the phenomena of interest occur below the electron plasma frequency ω_{pe} [72].

In the electrostatic approximation, the electric field is the variable in the system, which is then independent of the magnetic field. The curl of an electric field is zero, which writes

$$\nabla \times \boldsymbol{E} = 0. \tag{2.46}$$

Since the electric field becomes irrotational, it can be represented as a gradient of a scalar, called the electrostatic potential, in equation (2.3)

$$\boldsymbol{E} = -\nabla \Phi. \tag{2.47}$$

In addition, Gauss' law (2.2) is simplified. After introducing the potential, the so-called Poisson equation follows

$$-\nabla \cdot (\varepsilon_0 \nabla \Phi) = \rho, \tag{2.48}$$

and the radio-frequency form of generalized Ohm's law (2.45) can be expressed as

$$\frac{\partial \boldsymbol{j}}{\partial t} = -\varepsilon_0 \omega_{\mathrm{pe}}^2 \nabla \Phi - \nu \boldsymbol{j}. \tag{2.49}$$

Eventually, without induction effects, a consistent system of equations for the potential, charge density, and current density is given by the Poisson equation, the continuity equation, and the generalized Ohm's law. With the assumption of stationary ions $\boldsymbol{u}_{\mathrm{i}}(r) = 0$, definitions of the current density $\boldsymbol{j} = -en_{\mathrm{e}}\boldsymbol{u}_{\mathrm{e}}$ and space charge density $\rho = e\,(n_{\mathrm{e}} - n_{\mathrm{i}})$ can be found in Section 2.2.3. The reduced system of equations delivers good results under the assumption that no wave phenomena have to be taken into account that the relevant dynamics take place below the plasma frequency. Accordingly, the system can be solved with the consideration of suitable boundary conditions.

2.2.5 Collisions in Technical Plasmas

In general, two different fundamental collision mechanisms are considered in technical plasmas. As already mentioned, one type is the binary collisions between two particles, which in most cases dominate the energy losses due to the low ionization degree in technical plasmas; the other type is caused by the kinetic effects, which are interpreted as collisions. To be more specific, the electrons are deflected by the electric field and result in energy losses. Such phenomena are represented by a collision frequency. Notably, in a low-pressure regime, the kinetic effects become significant. Research [47] shows that the total collision frequency follows a linear relation with the rate of momentum transfer from electron to neutrals in addition to the kinetic effects. Consequently, it is convenient to assume that the effective collision frequency ν_{eff} is the sum of the elastic collision frequency ν_{col} and the kinetic collision frequency ν_{kin}.

In order to describe collision processes, K is defined as the rate constant, which is the collision frequency per unit density. With the neutral particle density n_{N}, the collision frequency in technical plasmas is

$$\nu_{\text{eff}} = n_{\text{N}} K. \tag{2.50}$$

Here, for an ideal gas, the neutral can be seen as the stationary background. The neutral particle density depends on pressure p and the gas temperature T_{N}. The corresponding relation writes

$$p = n_{\text{N}} K_B T_{\text{N}}. \tag{2.51}$$

The collision quantities in a plasma can be calculated from the integration over the velocity distribution functions of the particles:

$$\nu_{\text{eff}} = n_{\text{N}} \int \sigma(v_R) v_R f_1(\boldsymbol{v}_1) f_1(\boldsymbol{v}_2) \mathrm{d}^3 v_1 \mathrm{d}^3 v_2. \tag{2.52}$$

The rate constant in the integration form is then

$$K = \int \sigma(v_R) v_R f_1(\boldsymbol{v}_1) f_1(\boldsymbol{v}_2) \mathrm{d}^3 v_1 \mathrm{d}^3 v_2 \tag{2.53}$$

where $f_1(\boldsymbol{v}_1)$, $f_2(\boldsymbol{v}_2)$ are the distribution functions corresponding to velocity $\boldsymbol{v}_1$ and $\boldsymbol{v}_2$. The relative velocity is designated by $v_R = |\boldsymbol{v}_1 - \boldsymbol{v}_2|$, which describes the velocity difference between the particles before the collision. $\sigma(v_R)$ denotes the cross section, which depends on the velocity. For collisions between heavy particles with electrons, the velocity of the target particles is crucially smaller than that of the incident particles. Therefore, it is acceptable to assume $v_R = |\boldsymbol{v}_1|$. With such an assumption, the integration of $\boldsymbol{v}_2$ is trivially done.

As an initial approximation, isotropic Maxwellian distribution functions are used to simplify the integral, which is often the natural outcome of collisional processes. With given incident particle mass m and T, replacing v_1 by v, we obtain

$$K(T) = \left(\frac{m}{2\pi T}\right)^{3/2} \int_0^\infty \sigma(v) v \exp\left(-\frac{mv^2}{2T}\right) 4\pi v^2 \mathrm{d}^2 v. \qquad (2.54)$$

As is introduced in detail in [4], the elastic collisions between electrons and neutrals are dominant due to the low ionization degree in technical plasmas. Specifically, the collision frequencies for ionization and excitation are negligible since they are comparatively small. Therefore, the remaining term is electron-neutral collisions. To obtain the collision frequency, the cross section is required for solving the integral. It is sufficiently investigated for different gases in some literature [73].

$$\nu_{\mathrm{col}} = n_\mathrm{N} K_{\mathrm{col}}(T_\mathrm{e}). \qquad (2.55)$$

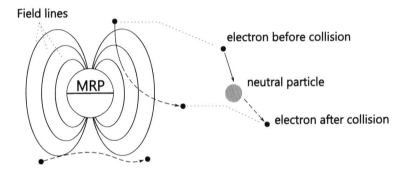

Figure 2.1: Illustration of the interaction between electrons and the probe: two different types of collisions are presented. On the left, the phenomena caused by kinetic effects are shown, while an elastic collision of an electron with a heavy neutral particle is sketched on the right .

Regarding the study of MRP, one primary problem concerning collisions is that the kinetic effects cannot be covered in the fluid model. In most cases, the energy losses are mainly related to the elastic electron-neutral collisions, where the kinetic effects can be neglected. However, in the low-pressure plasma, the impact of the kinetic effects becomes more pronounced. Briefly speaking, the particles are deflected by the field of the probe, and the energy is transported out of the perturbed domain of the plasma. As is shown in Figure 2.1, the electrons are influenced by the electric field lines of the idealized MRP, which results in a change in momentum.

Here, the idealization of MRP is discussed in Chapter 4, where the holder is neglected. When the electrons are at a farther distance, the MRP is not able to "see" those electrons to a certain extent. That is, some areas around the probe can be seen as the influenced domain, where the kinetic effects play an important role. Thus, together with the elastic collisions between electrons and neutral particles, the total energy losses can be described.

Since the kinetic effects are interpreted as collision frequency, it can be expressed with the thermal velocity and an undefined characteristic length L

$$\nu_{\mathrm{kin}} = \frac{v_{\mathrm{th,e}}}{L}, \tag{2.56}$$

it can be written as

$$\nu_{\mathrm{kin}} = \frac{\lambda_{\mathrm{D}}\omega_{\mathrm{pe}}}{L}, \tag{2.57}$$

where the Debye length λ_{D} depends on the electron temperature T_{e} in energy units,

$$v_{\mathrm{th,e}} = \sqrt{\frac{T_{\mathrm{e}}}{m_{\mathrm{e}}}}, \tag{2.58}$$

and the electron plasma frequency is related to the electron density n_{e},

$$\omega_{\mathrm{pe}} = \sqrt{\frac{e^2 n_{\mathrm{e}}}{\varepsilon_0 m_{\mathrm{e}}}}. \tag{2.59}$$

As defined previously, the effective collision frequency can be written as the sum of elastic collision frequency ν_{col} and kinetic collision frequency ν_{kin}

$$\nu_{\mathrm{eff}} = \nu_{\mathrm{col}} + \nu_{\mathrm{kin}}. \tag{2.60}$$

According to [74], the corresponding pressure can be converted directly from the elastic collision frequency depending on the investigated gas. However, the intensive study of the kinetic collision frequency remains unclarified. Especially, there has been no efficient model to determine the kinetic collision frequency for the pure collisionless cases.

Ultimately, as is introduced in equation (2.60), the total energy loss is directly related to the effective collision frequency, including elastic collision frequency and kinetic collision frequency. As previously stated, the energy loss mechanism in the fluid model is due to the elastic electron-neutral collisions, which means the energy transport is absent. However, in reality, the electrons can escape from the influenced domain and take away the energy, which is known as the kinetic effects. Especially in the low-pressure regime, the fluid model reaches its limitation since the kinetic effects cannot be described by it.

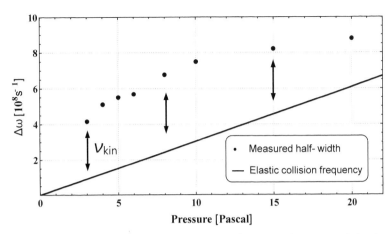

Figure 2.2: Comparison of the half-width $\Delta\omega$ between the measured spectra and the electron neutral collision frequency in the Drude model: the pronounced offset indicates the absence of the kinetic effects due to the electric field caused by the probe.

In Figure 2.2, the half-width is measured to compare with the results of the cold plasma model. The difference between the experimental data to the simulation results from the cold plasma model is observed. In this experiment, the half-width is measured for a plasma with electron density as $n_e = 3 \times 10^{16}\,\mathrm{m}^{-3}$, and electron temperature at $T_e = 3\,\mathrm{eV}$ in a cylindrical double inductively coupled plasma reactor. Although it is difficult to measure the energy loss at zero pressure, the offset can still clearly demonstrates the dominance of the kinetic collision frequency. As the pressure increases, the "gap" becomes smaller, which indicates the fluid model is more suitable for providing predictions of plasma at a relatively higher pressure. In this work, such kinetic effects are intensively investigated. To cover the limitation of the fluid model, kinetic approaches are necessary. In particular, numerical methods become a powerful tool to obtain the kinetic description of plasma, which is introduced in the following sections. Eventually, the relation between the electron temperature to the kinetic collision frequency is discussed. Based on the proposed spectral kinetic scheme, the corresponding simulation results are presented in Chapter 6.

2.2.6 Cold Plasma Model of MRP

This section is dedicated to the fluid description of the multipole resonance probe. The first investigation of the MRP is completed in the PluTO project by Martin Lapke in his dissertation, where the theoretical modeling of the MRP is exclusively dealt with [75]. Followed by Jens Oberrath's dissertation [76], the modeling and analysis of MRP with a functional analytical method are presented, where the limitation of the fluid description is discussed. In this work, the kinetic model with a numerical approach is intensively studied. Eventually, the comparison of the proposed spectral kinetic model to the cold plasma model of MRP is demonstrated. Therefore, the cold plasma model of MRP is briefly described here for completeness. The study of MRP based on the cold plasma model was well documented in [17, 47, 49].

The cold plasma model is derived in Section 2.2.3, where the dynamics of plasma are described by the equation of continuity (2.44) and the equation of motion of the electrons (2.49). In this system, the first equation expresses the conservation of charge, while the second equation covers the inertia of the electrons and momentum loss due to elastic electron-neutral collisions. It is often advantageous to utilize the linearity of the system for further simplification. By taking a time-harmonic approach, the quantities with time dependence can be expressed in the frequency domain. The plasma is to be described as a dielectric medium characterized by an equivalent permittivity. A linear relation between the current density j and the electric field E can be found.

Starting with the phasor notation, time derivatives simplify as $\partial/\partial t \rightarrow i\omega$. The complex electric fields can be written as:

$$E(r, t) = \underline{E}(r)\exp(i\omega t), \tag{2.61}$$

Under this condition, the simplified linear form of the continuity equation can be derived as

$$i\omega\rho = -\nabla \cdot j, \tag{2.62}$$

and the radio-frequency form of generalized Ohm's law

$$i\omega j = \varepsilon_0 \omega_{\text{pe}}^2 E - \nu j. \tag{2.63}$$

This allows an expressing of the so-called high-frequency conductivity of a plasma

$$\sigma_{\text{P}} = \frac{\varepsilon_0 \omega_{\text{pe}}^2}{i\omega + \nu}, \tag{2.64}$$

the relation between current density j and the electric field E yields

$$j = \sigma_{\text{P}} E. \tag{2.65}$$

Moreover, the plasma is treated as a dielectric medium. Thus, a displacement current is introduced as

$$\boldsymbol{D} = \varepsilon_{\mathrm{P}} \boldsymbol{E}, \tag{2.66}$$

where the so-called plasma permittivity ε_{P} can be determined from Ampere's law (2.5) in the time-harmonic case

$$\varepsilon_{\mathrm{P}} = \varepsilon_0 \left(1 - \frac{\omega_{\mathrm{pe}}^2}{\omega(\omega - i\nu)} \right). \tag{2.67}$$

Regarding the MRP, due to the breaking of the spherical symmetry caused by the holder, it is difficult to derive an analytical description of the MRP. Therefore, it is convenient to neglect the holder, which is justified as an appropriate assumption in [17]. The idealized MRP is then proposed for the analytical model. A detailed explanation of the concept of ideal MRP can be found in Section 4.1. As is shown in Figure 2.3, the ideal MRP holds a geometrical symmetry. The probe head consists of two metallic hemispheres encased in a dielectric layer. An isotropic plasma is assumed in the considered domain, where the MRP is immersed. Notably, an additional forming plasma sheath has to be considered around the probe within the plasma, which can be treated as a dielectric matter with a relative permittivity.

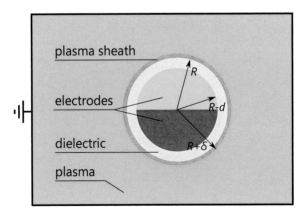

Figure 2.3: Schematic of the idealized MRP: the electrodes are covered by the dielectric matter with a plasma sheath considered within the plasma. The holder is neglected to provide symmetry geometrically.

Thus, for the electrostatic case, the system consists of a partial differential equation in the form of the momentum equation and a second-order differential equation in the form of the Poisson equation. The boundary conditions are given by the continuity equation. It is assumed that there are no electrons in the dielectric matter and sheath. Consequently, Poisson's equation gives the potential as

$$-\nabla\cdot(\varepsilon_0\varepsilon_r\nabla\Phi(\boldsymbol{r})) = \begin{cases} \rho, & \text{in the plasma} \\ \\ 0, & \text{in the dielectric and the sheath} \end{cases} \tag{2.68}$$

with the permittivity

$$\varepsilon_r = \begin{cases} 1, & \text{in the plasma and the sheath} \\ \\ \varepsilon_D, & \text{in the dielectric} \end{cases} . \tag{2.69}$$

The boundary condition at the surface of the electrodes is defined as the applied voltage, which writes

$$\Phi|_{r=R-d} = \begin{cases} +U(t), & \text{for } 0 \leq \theta < \pi/2 \\ \\ -U(t), & \text{for } \pi/2 < \theta \leq \pi \end{cases} , \tag{2.70}$$

and the vanishing of the potential at infinity is assumed,

$$\Phi|_{r\to\infty} = 0. \tag{2.71}$$

As previously described, it is often possible to take a time-harmonic approach. It reduces the system of equations to a single equation of the Laplace equation

$$-\nabla\cdot(\varepsilon_0\varepsilon_r\nabla\Phi(\boldsymbol{r})) = 0. \tag{2.72}$$

Depending on the medium, dielectric, sheath layer ,or plasma, different permittivity is considered. Such as the well-known Drude formula for the relative permittivity of plasma is derived in (2.67). The permittivity of dielectric regions is assumed as ε_D, and the sheath is treated as a dielectric layer with a relative permittivity identical to unity. Thus, the differential equation is with the permittivity as

$$\varepsilon_r = \begin{cases} \varepsilon_D, & \text{in the dielectric} \\ \\ 1, & \text{in the sheath} \\ \\ 1 - \dfrac{\omega_{pe}^2}{\omega(\omega - i\nu)}, & \text{in the plasma} \end{cases} . \tag{2.73}$$

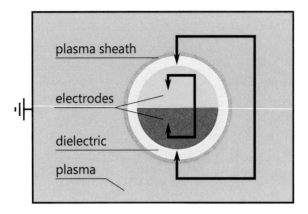

Figure 2.4: Schematic of the idealized MRP: the coupling occurs between the electrodes (through the dielectric and the plasma).

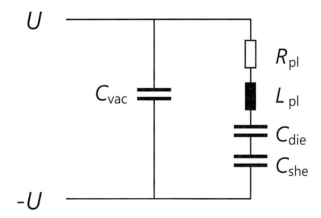

Figure 2.5: Equivalent lumped element circuit of the MRP: The capacitor C_{vac} refers to the vacuum coupling between the electrodes, the resistance R_{pl} represents the collisions within the plasma, the inductor L_{pl} describes the inertia of the particles, and the capacitor C_{she}, C_{die} relates to the plasma sheath and dielectric layer.

In [47], it is found that the behavior of the investigated model can be described by equivalent lumped element circuits. As shown in Figure 2.4, the vacuum coupling between the electrodes through the dielectric and the plasma are considered, whereas the coupling from the electrodes to the ground is neglected due to the vanishing of the field.

An illustration of the equivalent lumped element circuit is demonstrated in Figure 2.5. To be more specific, the MRP is symmetry with regard to the electrical behavior with respect to the mapping $\omega t \rightarrow \omega t + \pi$ and $\Phi \rightarrow -\Phi$. Therefore, the voltage applied on each electrode can be simply given as U and $-U$, respectively. The vacuum coupling directly between the electrodes is denoted by C_{vac}. The energy loss in the plasma caused by the elastic electron-neutral collisions is interpreted as a resistance R_{pl}. Besides, the inductor L_{pl} describes the mass inertia of the electrons, and the capacitors, C_{she} and C_{die}, describe the plasma sheath and dielectric matter. Here, these two capacitors can be combined for simplification.

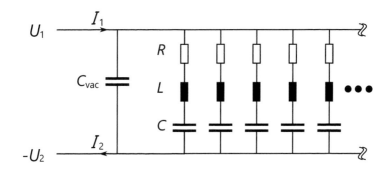

Figure 2.6: Equivalent lumped element circuit of the MRP: The capacitor C_{vac} denotes the vacuum coupling. Each resonance mode is represented by a series circuit. The resonance mode goes up to infinite. However, in practice, only a finite number of resonance modes are considered.

In reality, there is an infinite number of coupling between the electrodes, which can be treated as an infinite number of parallel series resonant circuits in parallel to a vacuum coupling. Here, each discrete resonance mode is then represented by a series circuit. It is noticeable that the coupling from the electrodes to the ground (or infinity) is not considered. It is due to the fact that the field decays rapidly with increasing distance from the ideal MRP. As has been demonstrated above, the equivalent resonant circuit simplifies to the structure presented in Figure 2.6 to describe the plasma dynamics. In practice, only a certain number of resonance modes are considered in the calculation.

In brief, a potential problem is solved in the Poisson equation. Eventually, with the given plasma parameters, the resonance response to the signal provided via the devices can be analytically determined. Since the expression of the potential can be found explicitly, a displacement current due to the dielectric can be directly obtained. Overall, the ratio of the calculated current to the applied voltage is derived as the system response in the frequency domain, which is known as the admittance of the system:

$$Y(\omega) = I/U = i\omega C_{\text{vac}} + \sum_{l=1}^{\infty} C_l \left(\frac{1}{i\omega} + \frac{i\omega + \nu}{\eta_l^2 \omega_{\text{pe}}^2} \right)^{-1}. \tag{2.74}$$

Here, the index l is the mode number. It is noticeable that the modes with an even number l are canceled out due to the antisymmetry of the ideal MRP. That is, only the odd l is considered, where $l = 1, 3, 5, , \ldots, \infty$. Besides the vacuum coupling C_{vac}, C_l describes the capacitances of the resonance circuits. These coefficients (such as C_{vac}, C_l, and η_l) depend on the parameters of the IMRP.

Specifically, the capacitance C_l is introduced as

$$C_l = \frac{a_l \varepsilon_0 R}{\left(1 + \frac{\delta}{R}\right)^{2l+1} - b_l}. \tag{2.75}$$

The resonance frequencies and the damping of the resonance behavior are in particular of great importance in this work. Then the resonance frequency is defined as

$$\omega_{\text{res},l} = \eta_l \omega_{\text{pe}}, \tag{2.76}$$

where the exact expression of the resonance frequencies $\omega_{\text{res},l}$

$$\omega_{\text{res},l} = \sqrt{\frac{l+1}{2l+1} \left(1 - b_l \left(1 + \frac{\delta}{R} \right)^{-2l+1} \right)} \cdot \omega_{\text{pe}}. \tag{2.77}$$

The terms a_l, b_l are dimensionless algebraic functions that consider the configuration parameters such as the radius of the idealized MRP and dielectric properties. The detail of the functions can be found in [75]. Another important unknown parameter is sheath thickness δ. In the cold plasma model, a sheath thickness is required to determine the resonance frequency. Apparently, the resonance frequency strongly depends on the sheath thickness. As is earlier assumed in [77], the sheath thickness in the model is given as $3\lambda_{\text{D}}$, which can be seen as a reasonable approximation.

In addition, the quality factors as a measure of the damping can be represented as

$$Q_l = \frac{\eta_l \omega_{\text{pe}}}{\nu}. \tag{2.78}$$

Figure 2.7 shows a typical spectrum of the real part of the admittance of the ideal MRP, which is linked to the dissipation in the plasma. As previously stated, each resonance mode is represented by a series resonant circuit. The complex admittance $Y(\omega)$ is evaluated as the plasma-dependent frequency response of the system. As an example, the plasma with some given parameters is therefore investigated. The ideal MRP is assumed with a total radius $R = 4\,\mathrm{mm}$, including the surrounding dielectric coating of $d = 1\,\mathrm{mm}$. The relative permittivity of the dielectric matter $\varepsilon_D = 4$ is considered. Besides, the sheath thickness is set as $\delta = 0.386\,\mathrm{mm}$, corresponding to the size of Debye length. Regarding the plasma parameters, the electron density is set equal to $1 \times 10^{16}\,\mathrm{m}^{-3}$, which indicates that the electron plasma frequency is obtained as $\omega_{\mathrm{pe}} = 5.655 \times 10^9\,\mathrm{s}^{-1}$. Importantly, since the energy loss is solely caused by electron-neutral collisions, no kinetic effects can be captured.

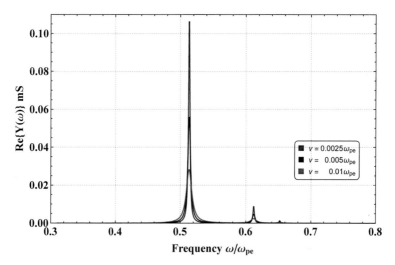

Figure 2.7: Resonance behavior of the ideal MRP based on the cold plasma model: The radius of the probe R is $4\,\mathrm{mm}$, the surrounding dielectric is with a thickness $d = 1\,\mathrm{mm}$, and the sheath thickness is assumed $\delta = 0.386\,\mathrm{mm}$. The relative permittivity of dielectric is given as $\varepsilon_D = 4$. Examples with different collision frequencies are presented.

In the presented simulation, the electron-neutral collision frequency ν is set to $0.0025\,\omega_{\mathrm{pe}}$, $0.005\,\omega_{\mathrm{pe}}$, and $0.01\,\omega_{\mathrm{pe}}$, respectively. The broadening of the resonance curves can be observed with the increase of the electron-neutral collision frequency. It indicates the relation between the collision frequency to the energy loss. In addition, since the sheath thickness is fixed as a constant, the resonance frequency does not vary in the comparison. Notably, the sheath thickness mostly depends on the electron density, which is discussed in Section 5.2.1. Therefore, the increase in the electron density leads to a higher resonance frequency. Apparently, the resonance frequency $\omega_{\mathrm{res},1} = 0.514\,\omega_{\mathrm{pe}}$ and $\omega_{\mathrm{res},3} = 0.612\,\omega_{\mathrm{pe}}$

can be clearly observed while the ones in higher modes are barely seen. That is, although several resonance peaks are demonstrated for different collision frequencies, the first one is dominant in the resonance structure. This dominating resonance peak is known as the dipole mode, where the coefficient is set as $l = 1$. In sum, as the response of the system, the dipole resonance is particularly of importance for the determination of the electron density and electron temperature.

In the aforementioned analytical model proposed by Lapke, the holder is neglected as an idealization to provide a geometrical symmetry. To verify the applicability of the idealization of MRP, the numerical simulation of the MRP is implemented within 3D-electromagnetic field simulations using CST Microwave Studio [48, 50, 51], which allows for the simulation of a complex model in the cold plasma approximation. In these studies, the complete geometry of the probe is considered in the description of the model, which consists of the probe head and holder. Within the CST Microwave Studio, modeling of the parameterized plasma efficiently determines the characterization of the behavior of the probe-plasma system with different setups. Regardlessly, a vastly similar resonance behavior is observed in the simulation results. By the investigation of the plasma and MRP with identical parameters, the comparable resonance peaks, in addition to the resonance frequencies, indicate the validity of the idealization of the MRP enormously. A good qualitative agreement is shown in the comparison between the analytical model and the numerical simulation [49].

Moreover, the numerical implementation of the simulation of the MRP-plasma system clearly confirms the significant importance of the dominating dipole resonance peak. The modes in higher order are therefore negligible in the calculations. That is, as shown in Figure 2.8, using only the dipole mode for the evaluation can be seen as an adequate approximation. The characterization of the plasma can be realized from the obtained resonance frequency and the full width at half maximum in the dipole resonance curve. Eventually, this unique feature of MRP leads to a remarkable optimization in the spectral kinetic model of the ideal MRP, which is discussed in detail in Section 4.3.

To demonstrate the kinetic effects, the comparison of the resonance response in dipole mode between the cold plasma model and the spectral kinetic model is discussed in the later chapter. Therefore, it is more practical to focus on the dipole mode solely, where the resonance frequency of it based on the cold plasma model is summarized as [19]

$$\omega_{\mathrm{res},1} = \sqrt{\frac{2}{3}\left(1 - 0.82\left(1 + \frac{\delta}{R}\right)^{-3}\right)}\,\omega_{\mathrm{pe}}, \tag{2.79}$$

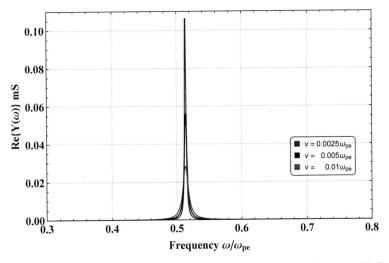

Figure 2.8: Dipole resonance behavior of the ideal MRP based on the cold plasma model: Same parameters for the probe and the plasma are investigated. Only $l = 1$ is considered in this case. Examples with different collision frequencies are presented.

3 Numerical Methods

As introduced in the previous chapter, different mathematical models are used to describe the various process occurring in plasmas. With given operating conditions, those models are expected to predict the plasma behavior and disclose the characterization of the investigated plasma. The fluid model is the most straightforward way of describing the plasma behavior, which treats the plasma properties as a function of macroscopic quantities. The fluid description, which solves the continuity equations for mass, momentum, and energy, is previously introduced. Specifically, the research of the MRP based on the cold plasma model has been intensively studied. However, the disadvantage of the fluid description of MRP becomes more pronounced inevitably since it is less suitable for low-pressure plasmas: the energy transport of the particles from the electromagnetic field is more influential than due to collisions with heavy species.

Therefore, the kinetic approach is necessary for the investigation of plasma in the low-pressure regime. As stated, kinetic models are built upon the Boltzmann equation, where a microscopic description of particles in the discharge is analyzed. Instead of as one continuum, all particles are computed individually in kinetic models. Compared to the fluid models, a more accurate explanation of the plasma behavior is provided by the kinetic models. Especially regarding the limitation of the fluid model of MRP, the energy loss due to the escape of the particles can be captured in the kinetic model. Basically, kinetic models are able to correctly predict the equilibrium or non-equilibrium microscopic behavior of particles at low pressure, including both ions and electrons.

It is known that the major drawback of the kinetic description of plasmas is the possible expensive computational cost. Apparently, considering a large number of particles in the plasma, obtaining numerical solutions to the full Boltzmann equation is a very complex problem. However, with the development of computer science for many years [78, 79], particle-based models are becoming the focus of research due to their excellent scalability. Moreover, many studies related to numerical methods and computer simulation of particles are documented [80,81,82,83,84,85,86], which provides the possibility to enhance the efficiency of the computation in the kinetic simulations. Therefore, kinetic models using some numerical methods are becoming powerful tools to tackle complicated problems. For example, particle-in-cell (PIC) is one of the well-known methods which is widely used in the modeling and simulation of plasmas. Using a particle-in-cell method, some of the expense of a full six-dimensional kinetic description can be reduced significantly. In PIC simulations, magnetic and electric fields are determined by solving Maxwell equations, then trajectories of individual particles are computed via the equation of motions. The unique feature of PIC simulation is the numerical grid cell in the space. Conceptual, the information of the magnetic and electric fields within the cell is stored at grid nodes. By interpolation and extrapolation, the communication between the fields and the particles is realized in such efficient simulation algorithms.

To reveal the kinetic effects in the investigation of MRP, a kinetic approach is proved to be necessary. For the simulation of the MRP-plasma model, PIC code is absolutely a promising candidate. The accurate physics is described by such numerical methods, where the missing energy loss in the cold plasma model can be understandably captured. Remarkably, the spectral kinetic method is proposed as a more straightforward numerical method in particular solving MRP problems. Taking advantage of the geometry of the ideal MRP, the spectral kinetic model is able to work without any numerical grids. The electrical and geometrical symmetry allows an efficient and reliable simulation of the IMRP system. The scheme is highly optimized mathematically since numerous interactions between particles are canceled due to symmetry. Furthermore, the assumption of using only the dipole mode simplifies the calculations to a large extent.

Additionally, the collisions between charged particles and neutrals play a vital role in the description of the behavior of plasma even at low pressure. The issue arises in particle-based simulations, where classical mechanics is not suitable for describing the elastic electron-neutral collisions since the number of particles involved is too large to be investigated. Hence, the time-honored Monte Carlo method, which was first introduced back in 1949 [87], is applied to handle such situations. The electron-neutral collisions are predicted probabilistically. Eventually, collisions in the IMRP-plasma system can be included by coupling the PIC method or the spectral kinetic method with a Monte Carlo collision (MCC) method.

The core of this dissertation is the kinetic simulation of IMRP. In this chapter, a brief discussion of the numerical methods is demonstrated, including the well-known particle-in-cell (PIC) method and the proposed spectral kinetic method for the modeling of IMRP, and the Monte Carlo collision method for the description of the collisions.

3.1 Super-Particles

The plasma kinetic models for the real physical systems are extremely extensive with respect to the number of interacting particles. Depending on the volume of the reactor and the particle density, the number of real particles in ideal plasma reaches up to 10^{12} to 10^{14}. It increases the complexity of the investigated particle model. Particularly, due to the capabilities and limitations of the computer, the computational cost can be excessively high in such kinetic simulations. Therefore, to make plasma simulations more efficient, some optimized simulation concepts are applied in the simulation process. One widely used concept is super-particles (SP), which is able to greatly reduce the number of simulated particles in the computational work.

Table 3.1: The influence of the super-particle on plasma parameters

Changed properties	Unchanged properties	
$q^s \to Wq$	$\rho^s \to \rho$	$\rho^s = m^s n_e^s = Wm \cdot \frac{1}{W} n_e$
$m^s \to Wm$	$v_{th}^s \to v_{th}$	$v_{th}^s = \sqrt{\frac{T_e^s}{m^s}} = \sqrt{\frac{WT_e}{Wm}}$
$T_e^s \to WT_e$	$\omega_{pe}^s \to \omega_{pe}$	$\omega_{pe}^s = \sqrt{\frac{e^s 2 n_e^s}{\varepsilon_0 m^s}} = \sqrt{\frac{W^2 e^2 \frac{1}{W} n_e}{\varepsilon_0 Wm}}$
$n_e^s \to \frac{1}{W} n_e$	$\lambda_D^s \to \lambda_D$	$\lambda_D^s = \sqrt{\frac{\varepsilon_0 T_e^s}{e^s 2 n_e^s}} = \sqrt{\frac{\varepsilon_0 WT_e}{W^2 e^2 \frac{1}{W} n_e}}$
$\Lambda^s \to \frac{1}{W} \Lambda$		

The idea of the concept of super-particles is to represent a certain number of real particles by some computational particles in the plasma models. Based on the fact that the Lorentz force depends only on the charge to mass ratio given by q/m, the number of particles in the simulation is rescaled. Here, the superscript "s" is referring to super-particles; the factor W between real particles N^{real} to super-particles N^s can be set around 10^5 to 10^6, which allows efficient simulations

$$N^{real} = WN^s. \tag{3.1}$$

Although the kinetic simulation can be significantly optimized in terms of efficiency, super-particles give rise to some problems which cause errors in the results. That is, some plasma parameters are scaled incorrectly under the assumption of super-particles, which are partly listed in Table 3.1. For instance, in plasma physics, a distinguishing feature is the presence of collective effects, which is denoted by plasma parameter Λ [88]. This dimensionless parameter relates to the number of particles in a Debye sphere with Debye length λ_D,

$$\Lambda = \frac{4\pi}{3} n_e \lambda_D^3. \tag{3.2}$$

It describes the collective status of plasma: a collection of charged particles can be seen as an ideal plasma when $\Lambda \gg 1$. However, the plasma parameter decreases excessively for super-particles due to the smaller electron density

$$\Lambda^{real} = W\Lambda^s. \tag{3.3}$$

Additionally, the Coulomb logarithm is introduced to classify the type of plasmas, which is designated by $\ln \Lambda$. In particular, a system is considered weakly coupled when $\ln \Lambda$ is large and strongly coupled when it is small. For typical ideal plasmas, the Coulomb logarithm lies in the range 10 to 20 for typical ideal plasmas [89]. Apparently, by using super-particles, the Coulomb logarithm tends to diminish close to zero, which breaks down the initial assumption. In such cases, it forms a very sparse and fragile type of solid, often referred to as a Coulomb crystal instead of ideal plasma.

Moreover, in the ideal plasma model, the frequency of electron-neutral collision is much larger than the frequency of electron-electron collision or electron-ion collision. Therefore, electron-electron collisions and electron-ion collisions are considered negligible in ideal plasma simulations. However, those trivial collisions become pronounced resulting from the effect of super-particles.

Taking electron-electron collision frequency as an example, we write

$$\nu_{\text{ee}} = v n_{\text{e}} \sigma_{\text{eff}} \tag{3.4}$$

where v is the velocity of electrons, n_{e} is the electron density, and σ_{eff} is the effective area of small angle collisions [90].

The effective area of small angle collisions is given as

$$\sigma_{\text{eff}} = \frac{e^4}{4\pi \epsilon_0^2 (2T_{\text{e}})^2} \ln \Lambda. \tag{3.5}$$

Thus, the electron-electron collision frequency is

$$\nu_{\text{ee}} = v n_{\text{e}} \frac{e^4}{4\pi \epsilon_0^2 (2T_{\text{e}})^2} \ln \Lambda. \tag{3.6}$$

For the cases with the assumption of super-particles, the discussed collision frequency increases significantly since some of the plasma parameters are changed, especially the factor W is a relatively large number. Thus, the collision frequency $\nu_{\text{ee}}^{\text{s}}$ for the plasma with super-particles is

$$\nu_{\text{ee}}^{\text{s}} \approx W^3 \nu_{\text{ee}}, \tag{3.7}$$

which leads to a drastic deviation in collision frequency.

Furthermore, the Coulomb interactions are of enormous importance in the kinetic model considering the fact that most of the computation relates to the interacting particles. However, they tend to become stronger between super-particles, especially within a short distance. Inevitably, those exaggerated coulomb forces intensively affect the characterization of the plasma in the kinetic simulations. Therefore, the drawback of the super-particles needs to be resolved by certain numerical techniques, which are discussed in Section 3.2 and Section 3.3.

3.2 Particle-in-Cell

With the development of computer science over the last decades, particle simulation has become a recognized tool in research due to its excellent performance in the evaluation of complex systems. One of the well-known computational techniques is the so-called "particle-in-cell" (PIC) method, which is the standard approach to the kinetic model. Back in the 1950s, this machine calculation method is first introduced by Harlow [80]. Then it gained more popularity for plasma simulation. In the late 1950s, the motion of hundreds of particles, including interactions between each particle, was successfully simulated in PIC by Buneman [81] and Dawson [82]. In fact, PIC simulations have brought tremendous new insights practically in all branches of plasma physics. In [84,85], the simulation of some diagnostic devices was presented using a method similar to PIC.

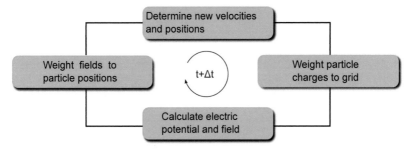

Figure 3.1: Scheme of the Particle-in-cell method: The dynamics of particles are calculated in the numerical cells and interpolated on the mash points, then the electric field is determined from the grid points in each iteration.

Nowadays, the PIC simulation represents a powerful tool that allows the fully kinetic description of the high-dimensional plasma. With the advanced computational devices and the improved programming techniques, it is possible to simulate the plasma behavior by following the trajectories of a large number of individual particles and incorporating complicated particle and plasma-surface interactions [83]. In a nutshell, PIC codes are usually robust and efficient. Generally speaking, a PIC code solves Maxwell equations with current and charge densities calculated from the simulated particles. Then the solutions for equations of motion with the Lorentz force for each individual particle are determined. The unique feature of this computational technique is that the newly computed magnetic and electric fields are stored on numerical grids, whereas the particles evolve continuously in position and velocity space. Hence the name of the method is "particle-in-cell". In Figure 3.1, a simplified scheme of PIC simulation is demonstrated. PIC simulation starts with a specified state of the system as initialization and ends with the output of the results. Over a time step Δt, the positions and the velocities of the particles can be calculated from the equations of motion. After the extrapolation of charge and current source to mesh points, the fields are computed in the so-called field solver. Then, the force acting on the particles is derived by interpolating fields at particle positions. The cycle is executed repeatedly until the conditional convergence is reached.

The particle-in-cell scheme can be simplified as follows:

- Computing r and v from the equations of motion (particle pusher).

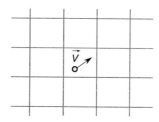

- Determining ρ from x and v (extrapolation: particles to grid).

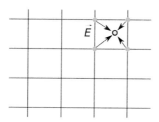

- Solving partial differential equations on mesh to obtain E field (field solver).

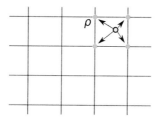

- Interpolating the fields E from the mesh to the particle locations r (interpolation: grid to particles).

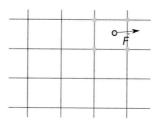

To optimize the particle-based model, the aforementioned concept of super-particles is often coupled in the PIC simulation to reduce the simulated particle number. In principle, noise or fluctuation is inevitably inherent to particle simulations. The possible issues in the kinetic simulation caused by the super-particles are discussed in Section 3.1: the exaggerated Coulomb interaction leads to severe errors. However, such errors can be resolved in the PIC simulation. That is, information between the super-particles within the cells and the fields stored at grid nodes is communicated via interpolation. Thus, the super-particles in the same cell are considered clouds of charge. It implies a vanishing two-point force between super-particles at short distances, which avoids exaggerated errors [86]. Besides, parallelization is applicable for large-scale PIC simulations. In summary, the PIC method is conceptually simple, and it can be efficiently implemented in a massively parallel framework.

In the following, the basic idea of the PIC scheme is presented. For a random particle p, the charge is given as q_p, and mass is m_p at position $\boldsymbol{x}_p$ with a velocity $\boldsymbol{v}_p$, then the Lorentz force acting on the particle is

$$\boldsymbol{F}_p = q_p \left(\boldsymbol{E}(\boldsymbol{r}_p) + \boldsymbol{v}_p \times \boldsymbol{B}_p(\boldsymbol{r}_p) \right). \tag{3.8}$$

To compute the force acting on the particles, the fields at the particle position are required. Those fields are created by the additional sources and the particles in the system, which can be evaluated from Maxwell's equations:

$$\nabla \cdot \boldsymbol{E} = \frac{\rho}{\varepsilon_0}, \tag{3.9}$$

$$\nabla \times \boldsymbol{E} = -\frac{\partial \boldsymbol{B}}{\partial t}, \tag{3.10}$$

$$\nabla \cdot \boldsymbol{B} = 0, \tag{3.11}$$

$$\nabla \times \boldsymbol{B} = \mu_0 \boldsymbol{j} + \mu_0 \varepsilon_0 \frac{\partial \boldsymbol{E}}{\partial t}. \tag{3.12}$$

Here, we assume the example is in the electrostatic approximation. Then the magnetic field is neglected. Thus, the electric field can be directly calculated in Poisson's equation (3.9). As a matter of fact, solving such partial differential equations is commonly known as the field solver in PIC simulations. In numerical plasma physics, many techniques, such as finite difference methods, finite element methods, and spectral methods, are developed to provide the numerical solutions reliably [91].

Moreover, the trajectory of all the super-particles is followed during a PIC simulation. This part is often known as particle pusher or particle mover, which is required to be accurate and efficient. It is usually associated with solving the equation of motion of particles with the Lorentz force. The motion of each super-particle is governed by Newton's Second Law

$$\frac{\mathrm{d}\boldsymbol{x}_p}{\mathrm{d}t} = \boldsymbol{v}_p, \tag{3.13}$$

$$m_p \frac{\mathrm{d}\boldsymbol{v}_p}{\mathrm{d}t} = \boldsymbol{F}_p. \tag{3.14}$$

The system of ordinary differential equations is to be solved. As introduced, the discretization can be realized by the finite differences method. Generally speaking, the major requirements for the particle pusher are high accuracy and speed. The most common and straightforward approach is based on the explicit leapfrog integration, which is known as a second-order symplectic method. In a nutshell, the leapfrog algorithm is able to provide the solutions with accuracy and stability at relatively low costs

$$\frac{\boldsymbol{v}(t + \frac{1}{2}\Delta t) - \boldsymbol{v}(t - \frac{1}{2}\Delta t)}{\Delta t} = \frac{q_p}{m_p} \boldsymbol{E}_p, \tag{3.15}$$

$$\frac{\boldsymbol{x}(t + \Delta t) - \boldsymbol{x}(t)}{\Delta t} = \boldsymbol{v}(t + \frac{1}{2}\Delta t). \tag{3.16}$$

The sketch of the leapfrog algorithm is shown in Figure 3.2: The position of a particle is pushed from t to $t + \Delta t$, while the velocity at half time step $t + \frac{1}{2}\Delta t$ is calculated. To start the simulation, the initial conditions are necessary, such as the velocity at $t = -\frac{1}{2}\Delta t$. Then the cycle is executed over time Δt iteratively.

In principle, the computation time is highly reduced due to the interpolation in PIC simulations. To be more specific, in the field solver, the fields are determined on the grid points, then interpolated back to the particles via the field weighting. Instead of calculating the field quantities on every super-particle, the information of the fields is obtained from the numerical grid cells. That is, the space-dependent properties, such as the charge density or the current, are computed on the grid [92].

In addition, one of the vital subjects concerning the numerical simulation is the noise and statistical fluctuations, which is related to accuracy and stability issues. All particle-based simulations are noisy in contrast to the fluid model. Being a numerical model, especially due to the discretization, the accuracy and stability of PIC strongly depend on the choice of several simulation parameters, such as the size of the time step and the spatial resolution of the numerical grid. The problem has been discussed theoretically in depth to reach a precise understanding [93,94]. Briefly speaking, in the case of electrostatic, the time step Δt, the grid size Δx, and the number of super-particles need to be well defined to avoid numerical artifacts, which include numerical heating, non-linear oscillations, or divergence of the computation. Firstly, the grid size must resolve the Debye length λ_{D}

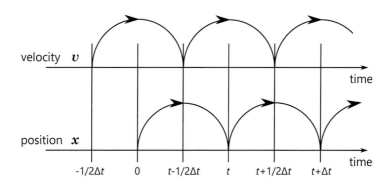

Figure 3.2: Illustration of the leapfrog algorithm

to prevent the improper description of the physics and finite grid instability. In fact, the lack of spatial resolution leads to severe incorrect physical phenomena. Therefore, it is necessary to at least require $\Delta x < \lambda_D$. Actually, a more rigorous analysis suggests the following criterion [86, 91]:

$$\Delta x < 3.4\lambda_D. \tag{3.17}$$

Secondly, the accuracy of the simulated trajectories of the particles are related to the size of the time step. That is, the time step of solving the equations of motion is required to be small to prevent the nonphysical trajectories. Since the electrons are quick to react to the fields in the simulations, the characteristic time scale in the plasma is given by the inverse electron plasma frequency ω_{pe}^{-1}. In order to ensure the stability of the solution, the criterion is given as

$$\Delta t < 2\omega_{pe}^{-1} \tag{3.18}$$

Typically, to guarantee the reliable operation of the simulation, a much more restrictive condition is used during PIC simulations

$$\Delta t < 0.2\omega_{pe}^{-1} \tag{3.19}$$

Lastly, as is introduced in 3.1, super-particles may cause errors in the PIC simulation. The numerical heating appears due to statistical fluctuations when the number of super-particles is too low. In order to minimize the discrete particle noise, it is necessary to have a sufficient number of super-particles in the simulation. Specifically, the ratio of the number of super-particles to the number of grid cells must be much greater than 1.

3.3 Spectral Kinetic Scheme

Although PIC is widely used for plasma modeling, full PIC codes are computationally expensive and may cause statistical noise. Thus, the spectral kinetic method is developed to efficiently capture the kinetic effects. Particularly, the plasma-probe system can be described as a collection of free particles without employing any numerical grids. In fact, the field or particle exhibits many random interactions in the system. They tend to cancel each other out due to the spherical symmetry of the investigated IMRP model. It allows the simplification of the mathematical model.

In Figure 3.3, the simulation scheme is shown: only particle pusher and field solver is considered in the spectral kinetic model [8]. The electric field or the potential can be calculated in the field solver, and the canonical equations of motion play the role of particle pusher. The dynamics of the self-consistent system are described in the Hamiltonian formalism, and the Poisson problem can be solved with a Green's function. Iteratively, it determines the electric field at each particle position, followed by the movement of the particles due to the newly calculated electric field.

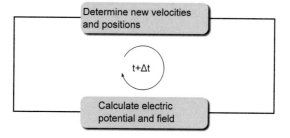

Figure 3.3: Scheme of the spectral kinetic simulation: The particles move individually according to the results of the dynamic equations in each iteration. The electric field is determined in the field solver, whereas the particle pusher gives the newly calculated position of particles.

As previously discussed, the errors caused by the super-particles, such as the exaggerated Coulomb interactions, are resolved in PIC simulations. The particles communicate via interpolation and extrapolation on the numerical meshes, which results in the vanishing of the particle interaction at short distances. However, such numerical grid cells are not defined in the spectral kinetic scheme. Therefore, an alternative approach is in particular required in the proposed model. As is introduced, the dynamics of the self-consistent system are described in the Hamiltonian formalism, and the Poisson problem can be solved with a Green's function. To be more specific, the explicit form of the Green's function is expressed in infinite series expansions as the formal solution. In the spectral kinetic scheme, the regularization to truncate the expansions is executed so that the singularity is avoided. Eventually, the simplified field solver can be obtained. The description of the system in Hamiltonian, in addition to the truncation of the sum of Green's expansions, is presented in detail in Chapter 4.

Regarding the particle pusher in the spectral kinetic scheme, one of the well-known drift-free higher-order algorithms is commonly attributed to Verlet. In 1967, the Verlet integration was contributed by Loup Verlet for use in molecular dynamics [78]. In this work, the velocity Verlet algorithm is employed in the particle pusher. In fact, related to the Verlet integration, the velocity Verlet algorithm was first introduced in 1982 [95]. The idea of this scheme is sketched in Figure 3.4. Similar to the leapfrog algorithm, the velocity and the position can be explicitly solved. However, the velocity and position are calculated at same time step.

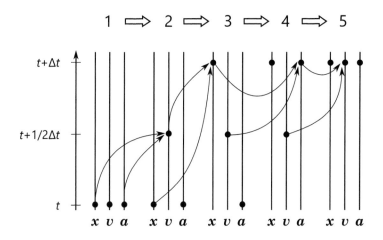

Figure 3.4: Illustration of the velocity Verlet integration scheme.

In a nutshell, there are five explicit steps in the velocity Verlet integration scheme in each iteration. It is assumed that a random particle locates at the starting position x_0, and it is with the velocity v_0 and the acceleration a_0. Thus, the algorithm can be summarized as follows:

1. Start the cycle according to the initial setup:

$$\boldsymbol{x}(t) = \boldsymbol{x}_0, \ \boldsymbol{v}(t) = \boldsymbol{v}_0, \ \boldsymbol{a}(t) = \boldsymbol{a}_0 \tag{3.20}$$

2. Calculate the velocity by half time step:

$$\boldsymbol{v}(t + \frac{1}{2}\Delta t) = \boldsymbol{v}(t) + \frac{1}{2}\boldsymbol{a}(t)\Delta t \tag{3.21}$$

3. Calculate new position:

$$\boldsymbol{x}(t + \Delta t) = \boldsymbol{x}(t) + \boldsymbol{v}(t + \frac{1}{2}\Delta t)\Delta t \tag{3.22}$$

4. Compute the acceleration from the interaction potential using $\boldsymbol{x}(t + \Delta t)$

$$\boldsymbol{a}(t + \Delta t) = f(\boldsymbol{x}(t + \Delta t)) \tag{3.23}$$

5. Calculate new velocity:

$$\boldsymbol{v}(t + \Delta t) = \boldsymbol{v}(t) + \frac{1}{2}\boldsymbol{a}(t + \Delta t)\Delta t \tag{3.24}$$

The unique feature of the spectral kinetic scheme compared to PIC is that the proposed method is completely grid-free. Therefore, unlike in the PIC simulations, there is no necessity for interpolation or extrapolation in each iteration of the calculation. As stated, for numerical simulations, it is a common issue that the stability condition needs to be ensured. In PIC, the criteria of the spatial resolution and the size of the time step are listed in Section 3.2. Since there are no numerical grids in physical space, restriction of spatial resolution can not be directly specified. However, a suitable criterion of the size of the time step restricts the distance of the traveling particles, which can avoid nonphysical trajectories in the simulation. The general criteria of the size of time step in PIC are derived (3.18), and often a much more restrictive condition is suggested (3.19). Similarly, in the particle pusher of the spectral kinetic model, the time step assumed in the velocity Verlet scheme is

$$\Delta t = 0.1 \, \omega_{\text{pe}}^{-1} \tag{3.25}$$

3.4 Monte Carlo Collision Method

The last missing part in the afore presented numerical schemes is the description of collision processes. The fact is, even in the low-pressure regime, the physics of energy transport
in elastic electron-neutral collisions is essential. The mass difference between the electrons
and ions allows that the ions can often be seen as stationary background, which is the
reason that more attention is paid to the collisions related to the electrons. However,
the particle-in-cell method and the spectral kinetic method are collisionless models. The
module within the introduced methods to describe the collision is coupled with the Monte
Carlo collision method.

As shown in Figure 3.5, the Monte Carlo collision model can be integrated into the
spectral kinetic scheme; the collision between particles is then evaluated at each iteration
in the simulation. Generally speaking, if the collisions are investigated with classical
mechanics in such particle simulations, the computational cost can be overwhelmingly
high. Hence, alternatively, the evaluation of the collision processes can be accomplished
probabilistically. The conventional Monte Carlo model is widely used in statistical physics,
which estimates an unknown value by using the principles of inferential statistics. It is
able to scheme the movement of the charged particles using random numbers [96]. To be
compatible with the proposed spectral kinetic method, where all the particle trajectories
are integrated simultaneously at the same time step, the Monte Carlo collision method
is developed with the consideration of a constant time step, including the null-collision
method [92, 97, 98].

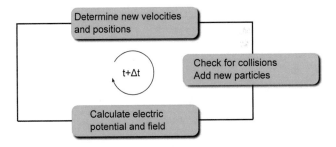

Figure 3.5: Scheme of the modified spectral kinetic simulation: The module of collision is integrated via the Monte Carlo collision model. The particles move individually according to the
results of the dynamic equations in each iteration. After the particle pusher gives the newly
calculated position of particles, then the collisions are evaluated. The iteration ends with a new
electric field calculated by the field solver.

In the kinetic simulation, at each time step in the iteration of the scheme, the occurrence of collisions is considered, which is expected to be consistent with the applied velocity Verlet algorithm. The assumption that the heavy neutrals are stationary is applied in electron-neutral collisions. As stated previously, the collision problem is given as a probabilistic interpretation. Mathematically, the underlying model starts from the probability distribution specified by the exponential probability density [99]

$$f(x) = \alpha \exp(-\alpha x). \tag{3.26}$$

Here, the corresponding exponential probability distribution function is obtained as

$$F(x) = 1 - \exp(-\alpha x). \tag{3.27}$$

for the coefficients $\alpha > 0$ and $x \geq 0$.

In the proposed model, the size of the time step in the scheme is given as a constant Δt. Then collision is assumed to take place in the time interval $[t, t + \Delta t]$. Defined by the exponential probability distribution (3.27), for a physical particle, the probability of the collision interaction within the time interval Δt can be specified from the probability of no occurrence of a collision

$$\begin{aligned}
P(t \leq \Delta t) &= 1 - P(t \geq \Delta t) \tag{3.28} \\
&= 1 - [F(t \to \infty) - F(\Delta t)] \\
&= 1 - \exp(-\nu \Delta t),
\end{aligned}$$

it is noticeable that x is replaced by t as it is in the time domain, and α refers to collision frequency, which is represented by ν. This collision frequency is related to the collision cross section σ and the neutral density. Since the background is assumed stationary, the relative velocity is directly obtained as v,

$$\nu = v n_{\mathrm{N}} \sigma. \tag{3.29}$$

Then a uniform random number in the range between 0 and 1, which writes $R_f \in [0, 1]$. This parameter is defined to compare with the obtained probability of a collision. Eventually, the condition to predict the collision interaction is written as

$$R_f \leq P(t \leq \Delta t) \leq 1, \tag{3.30}$$

where the occurrence of collision is decided within the specific time step Δt. That is, when the random number R_f is smaller than the probability, a collision will happen in the simulation. Thus, the particle is influenced by the collisional interaction. The new speed or direction of the particle can be updated. It is important to keep $P(t \leq \Delta t) \leq 1$. Moreover, the time step Δt must be small enough, which is verified in equation (3.25).

4 Spectral Kinetic Model

In this chapter, a spectral kinetic scheme of the idealized MRP is devised. The spherically symmetric geometry of IMRP simplifies the calculations for solving the kinetic model. Specifically, the dynamics of the self-consistent system are described in the Hamiltonian formalism, and the Poisson problem can be solved explicitly with a Green's function. Then, a certain truncation of infinite series expansions of the Green's function leads to an optimized mathematical model. Eventually, A simulation model of IMRP-plasma is established, which is with the applied voltage as input and the charge on the electrode as the response. This chapter is reproduced from [8] with the permission of AIP Publishing.

4.1 Idealized Multipole Resonance Probe

For this study, we strive to establish a dynamic model of the interaction of an active plasma resonance probe, the MRP, with a plasma. For the particle-based model, although the actual physical system can be accurately described by the finite ensemble of particles, the major challenge is the expensive computation due to the large number of particles in the simulation. To enhance the efficiency, some optimization can be applied, such as the aforementioned concept of super-particles and code parallelizing. Besides, permanently evolving computational devices make particle simulation very attractive.

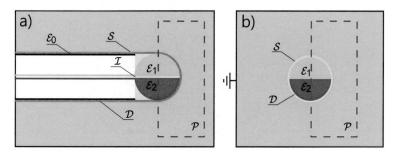

Figure 4.1: Schematic of a) the realistic MRP, and b) the idealized MRP: The highlighted hemisphere of the realistic MRP is treated as the ideal model, then the MRP can be idealized by the assumption of symmetry.

Another effective approach is the optimization of the modeling, which often utilizes the symmetry of the geometry to simplify the mathematical description of the investigated model. In the previous study of the MRP [21,22,51], the influence of different parameters of a modeled plasma is discussed to optimize the design of the device. The realistic MRP is investigated in 3D electromagnetic field simulations using CST Microwave Studio. The simulation allows a parameterized plasma-probe system, and it provides the options to study the characterization of the behavior virtually. As shown in Figure 4.1 (a), in domain $\mathcal{V}$, the probe is assumed to be coverd with dielectric $\mathcal{D}$ in the plasma bulk $\mathcal{P}$. The electrodes are ideal (infinite conductivity), which are separated from each other by ideal insulators $\mathcal{I}$ with vanishing permittivity and conductivity, and grounded surfaces are treated as an additional electrode $\mathcal{E}_0$. Notably, according to the simulated magnitude of the electric field for the MRP, the interaction occurs mostly in the highlighted area, which can be seen as an ideal model for the theoretical investigation. Moreover, in Figure 4.1 (b), the other half of the sphere is assumed symmetric. Then the IMRP can be formed with geometric symmetry.

The feasibility of the idealized MRP model was firstly discussed in [17]. The simple geometry of the idealized MRP allows an analytical evaluation of the measured signal, which is under the assumption of the electrostatic approximation. In [49], a full electromagnetic model of the realistic probe system was implemented in CST simulation. A similar resonance structure can be observed in both approaches. Thus, the assumption of the idealization of the MRP is proved to be applicable.

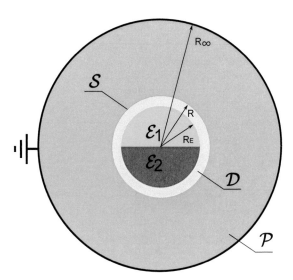

Figure 4.2: Schematic of the ideal axially symmetrical model of the MRP inside a plasma: the holder is neglected, which is suitable for the theoretical investigation.

In Figure 4.2, the IMRP of this work is illustrated: The chamber is given as a spatially bounded domain $\mathcal{V}$, and IMRP, which consists of two ideal electrodes, is immersed in the plasma volume $\mathcal{P}$: The ideal electrodes $\mathcal{E}_1$ and $\mathcal{E}_2$ with infinite conductivity are dielectrically covered in $\mathcal{D}$. The voltages u_1 and u_2 can be applied to the electrodes, which excite the system. The radius of the IMRP is R, and the thickness of the dielectric is Rd, which gives the radius of the electrodes $R_E = R - Rd$. The plasma around the probe is treated as an ensemble of N classical free point charges. The charge density $\rho(\boldsymbol{r})$ is given, including the constant surface charge density $\sigma_{\mathcal{S}}$, which indicates the total surface charge $Q_{\mathcal{S}}$ is homogeneously distributed on $\mathcal{S}$. The electrostatic approximation $\boldsymbol{E} = -\nabla\Phi$ is adopted in $\mathcal{V}$. The skin effect is negligible because the length scale is small compared to the skin depth, and no electromagnetic waves are emitted due to the fact that the frequency of the applied signal ω_{rf} is smaller than the electron plasma frequency ω_{pe} [100].

In the IMRP-plasma system, the electric potential can be calculated in Poisson's equation, where index n (or n') $= 1, 2$ refers to the corresponding electrode

$$-\nabla \cdot (\varepsilon_{\mathrm{r}}(\boldsymbol{r})\nabla\Phi(\boldsymbol{r})) = \rho(\boldsymbol{r}), \tag{4.1}$$

with boundary condition

$$\Phi(\boldsymbol{r}) = \begin{cases} 0, & |\boldsymbol{r}| \to \infty \\[2mm] u_n, & \boldsymbol{r} \in \mathcal{E}_n \end{cases}. \tag{4.2}$$

A suitable tool for the formal description of these relations is the Green's function $G(\boldsymbol{r}, \boldsymbol{r}')$, which gives

$$-\nabla \cdot (\varepsilon_{\mathrm{r}}(\boldsymbol{r})\nabla G(\boldsymbol{r}, \boldsymbol{r}')) = \delta^{(3)}(\boldsymbol{r} - \boldsymbol{r}'), \tag{4.3}$$

with

$$G(\boldsymbol{r}, \boldsymbol{r}') = 0, \quad \boldsymbol{r} \in \mathcal{E} \quad \text{or} \quad |\boldsymbol{r}| \to \infty \tag{4.4}$$

Green's Identities: The vector field $\boldsymbol{A}$ is assumed in three-dimensional space V with piecewise smooth boundary S indicated by ∂V. The divergence theorem states that

$$\int_V (\nabla \cdot \boldsymbol{A})\mathrm{d}V = \oint_{\partial V} \boldsymbol{A} \cdot \boldsymbol{n}\mathrm{d}S, \tag{4.5}$$

where $\boldsymbol{n}$ is the outward pointing unit vector normal to the surface.

For any arbitrary scalar functions ϕ and φ, let $\boldsymbol{A} = \phi\nabla\varphi$. Green's first identity is derived as

$$\int_V (\phi\nabla^2\varphi + \nabla\varphi \cdot \nabla\phi)\mathrm{d}V = \oint_{\partial V} \phi\nabla\varphi \cdot \mathrm{d}\boldsymbol{S}, \tag{4.6}$$

Then insert $\boldsymbol{A} = \phi\nabla\varphi - \varphi\nabla\phi$ into the divergence theorem, Green's second identity can be written as

$$\int_V (\phi\nabla^2\varphi - \varphi\nabla^2\phi)\mathrm{d}V = \oint_{\partial V} (\phi\nabla\varphi - \varphi\nabla\phi) \cdot \mathrm{d}\boldsymbol{S}. \tag{4.7}$$

In order to determine the solution of Poisson's equation (4.1), we assume

$$\begin{cases} \phi = \Phi(\boldsymbol{r}) \\ \varphi = G(\boldsymbol{r}, \boldsymbol{r}') \end{cases}, \tag{4.8}$$

which can be applied into Green's second identity (4.7).

Then the formal solution $\Phi(\boldsymbol{r})$ of the Poisson equation can be established. According to the boundary conditions, the Green's function vanishes at the electrodes and infinity. Notably, the normal vector of $\partial\mathcal{V}$ points outward of the domain, whereas the normal vector of the electrodes $\mathcal{E}_n$ points into the plasma domain. Therefore, the contribution of the electrodes is determined, where $\mathcal{E}$ represents all electrodes. Considering the interaction of the particles and the influence of $\sigma_{\mathcal{S}}$, the potential is

$$\Phi(\boldsymbol{r}, t) = \frac{1}{\varepsilon_0} \sum_{i=1}^{N} q_i G(\boldsymbol{r}, \boldsymbol{r}_i(t)) + \frac{1}{\varepsilon_0} \int_{\mathcal{S}} \sigma_{\mathcal{S}} G(\boldsymbol{r}, \boldsymbol{r}') \mathrm{d}f'$$
$$+ \int_{\mathcal{E}} \Phi(\boldsymbol{r}')\varepsilon_r(\boldsymbol{r}')\nabla'G(\boldsymbol{r}, \boldsymbol{r}') \cdot \mathrm{d}\boldsymbol{f}'. \tag{4.9}$$

Containing the information about the geometry, the latter formula is defined as the characteristic function $\Psi_n(\boldsymbol{r})$, which is independent of the plasma

$$\Psi_n(\boldsymbol{r}) = \int_{\mathcal{E}} \varepsilon_{\mathrm{r}}(\boldsymbol{r}')\nabla'G(\boldsymbol{r}, \boldsymbol{r}') \cdot \mathrm{d}\boldsymbol{f}'. \tag{4.10}$$

The characteristic function is the solution of Laplace equation

$$-\nabla\cdot(\varepsilon_{\mathrm{r}}(\boldsymbol{r})\nabla\Psi_n(\boldsymbol{r})) = 0, \tag{4.11}$$

with

$$\Psi_n(\boldsymbol{r}) = \begin{cases} 0, & |\boldsymbol{r}| \to \infty \\ \delta_{n'n}, & \boldsymbol{r} \in \mathcal{E}_{n'} \end{cases}. \tag{4.12}$$

Then the electric potential $\Phi(\boldsymbol{r}, t)$ inside $\mathcal{V}$ can be written as the interaction between a pair of particles in addition to the reaction of a particle to the applied voltages and surface charge $\Phi_{\mathcal{S}}$,

$$\Phi(\boldsymbol{r}, t) = \frac{1}{\varepsilon_0} \sum_{i=1}^{N} q_i G(\boldsymbol{r}, \boldsymbol{r}_i(t)) + \Phi_{\mathcal{S}}(\boldsymbol{r}) + \sum_{n=1}^{2} u_n(t)\Psi_n(\boldsymbol{r}). \tag{4.13}$$

The dynamics of plasma particles can be described by Hamiltonian, where the kinetic energy is obtained from the conjugate momentum $\boldsymbol{p}_k$ of free point charges with mass m_k to the position $\boldsymbol{r}_k$

$$H(\boldsymbol{r}_1, \ldots, \boldsymbol{r}_N, \boldsymbol{p}_1, \ldots, \boldsymbol{p}_N) = \sum_{k=1}^{N} \frac{\boldsymbol{p}_k^2}{2m_k} + V(\boldsymbol{r}_1, \ldots, \boldsymbol{r}_N). \tag{4.14}$$

The potential energy is determined from the electric potential in equation (4.13)

$$V(\boldsymbol{r}_1, \ldots, \boldsymbol{r}_N) = \frac{1}{\varepsilon_0} \sum_{k,i \in pair} q_k q_i G(\boldsymbol{r}_k, \boldsymbol{r}_i) + \sum_{k=1}^{N} q_k \left(\Phi_{\mathcal{S}}(\boldsymbol{r}_k) + \sum_{n=1}^{2} u_n(t) \Psi_n(\boldsymbol{r}_k) \right), \tag{4.15}$$

where *pair* is referring to the two-body interaction between physical particles k and i. Considering that each set of pair only interacts once, it can be written as

$$V(\boldsymbol{r}_1, \ldots, \boldsymbol{r}_N) = \frac{1}{2\varepsilon_0} \sum_{k=1}^{N} \sum_{\substack{i=1 \\ i \neq k}}^{N} q_k q_i G(\boldsymbol{r}_k, \boldsymbol{r}_i) + \sum_{k=1}^{N} q_k \left(\Phi_{\mathcal{S}}(\boldsymbol{r}_k) + \sum_{n=1}^{2} u_n(t) \Psi_n(\boldsymbol{r}_k) \right). \tag{4.16}$$

The proposed scheme consists of two modules, a field solver and a particle pusher. The electric field can be calculated as the field solver, and the canonical equations of motion play the role of particle pusher. The terms with the Green's function can be combined due to symmetry. Therefore, for a random particle k

$$\frac{\mathrm{d}\boldsymbol{r}_k}{\mathrm{d}t} = \frac{\partial H}{\partial \boldsymbol{p}_k} = \frac{\boldsymbol{p}_k}{m_k}, \tag{4.17}$$

$$\frac{\mathrm{d}\boldsymbol{p}_k}{\mathrm{d}t} = -\frac{\partial H}{\partial \boldsymbol{r}_k} = -\frac{1}{\varepsilon_0} \sum_{\substack{i=1 \\ i \neq k}}^{N} q_k q_i \nabla_k G(\boldsymbol{r}_k, \boldsymbol{r}_i) - q_k \nabla_k \Phi_{\mathcal{S}}(\boldsymbol{r}_k) - \sum_{n=1}^{2} u_n(t) \nabla_k \Psi_n(\boldsymbol{r}_k).$$

In general, a Green's function can be expanded in a set of basis functions depending on the geometry of the model. For the IMRP, the expansion is an infinite series of spherical harmonics with indices l $(0 \leq l < \infty)$ and m $(-l \leq m \leq l)$. In Table 4.1, the first few spherical harmonics in spherical coordinates are presented. (The detailed discussion is in the succeeding Section 4.2):

$$G(r, \theta, \varphi, r', \theta', \varphi') = \sum_{l=0}^{\infty} \sum_{m=-l}^{l} R_l(r, r') Y_{lm}^{\star}(\theta', \varphi') Y_{lm}(\theta, \varphi). \tag{4.18}$$

To avoid the singularity when two particles are almost identical, which leads to an infinite amount of energy, some regularization to truncate the noise in this model is required. In fact, the field or particle exhibits many random interactions in the system. They tend to cancel each other out due to the spherical symmetry, which allows the simplification of

Table 4.1: The first few spherical harmonics in spherical coordinates

l	m	$Y_{lm}(\theta, \varphi)$
0	0	$(1/4\pi)^{1/2}$
1	0	$(3/4\pi)^{1/2} \cos\theta$
1	± 1	$\mp(3/8\pi)^{1/2} \sin\theta e^{\pm i\varphi}$
2	0	$(5/16\pi)^{1/2}(3\cos^2\theta - 1)$
2	± 1	$\mp(15/8\pi)^{1/2} \sin\theta \cos\theta e^{\pm i\varphi}$
2	± 2	$\mp(15/32\pi)^{1/2} \sin^2\theta e^{\pm 2i\varphi}$
3	0	$(7/16\pi)^{1/2}(5\cos^3\theta - 3\cos\theta)$
3	± 1	$\mp(21/64\pi)^{1/2} \sin\theta(5\cos^2\theta - 1)e^{\pm i\varphi}$
3	± 2	$\mp(105/32\pi)^{1/2} \sin^2\theta \cos\theta e^{\pm 2i\varphi}$
3	± 3	$\mp(35/64\pi)^{1/2} \sin^3\theta e^{\pm 3i\varphi}$

the sum of Green's functions. The truncation is realized by a projection operator $\hat{\mathrm{P}}$ on subspaces

$$\frac{\partial H}{\partial \boldsymbol{p}_k} = \frac{\boldsymbol{p}_k}{m_k}, \tag{4.19}$$

$$-\frac{\partial H}{\partial \boldsymbol{r}_k} = -\hat{\mathrm{P}}\left(\frac{1}{\varepsilon_0}\sum_{\substack{i=1\\i\neq k}}^{N} q_k q_i \nabla_k G(\boldsymbol{r}_k, \boldsymbol{r}_i) + q_k \nabla_k \Phi_{\mathcal{S}}(\boldsymbol{r}_k) + \sum_{n=1}^{2} u_n(t)\nabla_k \Psi_n(\boldsymbol{r}_k)\right).$$

Due to the linearity, it is mathematically correct to shift the projection operator. Hence, the truncated Green's function can be found to effectively simplify the calculations, which is presented in Section 4.3.

$$\frac{\partial H}{\partial \boldsymbol{p}_k} = \frac{\boldsymbol{p}_k}{m_k}, \tag{4.20}$$

$$-\frac{\partial H}{\partial \boldsymbol{r}_k} = -\frac{1}{\varepsilon_0}\sum_{\substack{i=1\\i\neq k}}^{N} q_k q_i \nabla_k \hat{\mathrm{P}} G(\boldsymbol{r}_k, \boldsymbol{r}_i) - q_k \nabla_k \hat{\mathrm{P}} \Phi_{\mathcal{S}}(\boldsymbol{r}_k) - \sum_{n=1}^{2} u_n(t)\nabla_k \hat{\mathrm{P}} \Psi_n(\boldsymbol{r}_k).$$

In the kinetic scheme, the applied voltages at the electrodes are provided as the input of the simulation. As a response, the charges on the electrodes Q_n can be determined according to Gauss's law

$$Q_n = -\int_{\mathcal{E}_n} \varepsilon_0 \varepsilon_{\mathrm{r}}(\boldsymbol{r})\nabla\Phi(\boldsymbol{r}) \cdot \mathrm{d}\boldsymbol{f}. \tag{4.21}$$

4.2 Green's Function for the Ideal MRP

In electrostatics, the Green's function $G(\boldsymbol{r}, \boldsymbol{r}')$ is the solution of Poisson's equation with specified boundary conditions for a unit charge at the point $\boldsymbol{r}'$. For the case of the ideal MRP, the problem is to find the potential outside of a grounded sphere with a radius R_E that is covered with a dielectric of thickness Rd and permittivity ε_r so that the total device radius is $R = R_E + Rd$. Although charges only exist in the plasma bulk ($R \leq |\boldsymbol{r}'| < +\infty$), to provide a comprehensive understanding of the IMRP-plasma model, a Green's function of the full system is derived in this section, where $\boldsymbol{r}$, $\boldsymbol{r}' \in (R_E, +\infty)$ is assumed. It is defined that the sphere is grounded, and the potential vanishes at infinity, which is expressed in Poisson's equation (4.1). Then the jump condition for the discontinuity at $|\boldsymbol{r}| = R$ is considered to determine the solution in the dielectric and the plasma bulk.

The function $\varepsilon_r(\boldsymbol{r})$ describes the dielectric cover of the probe,

$$\varepsilon_r(\boldsymbol{r}) = \begin{cases} \varepsilon_r, & R_E < |\boldsymbol{r}| < R, \\ 1, & R < |\boldsymbol{r}| < +\infty, \end{cases} \tag{4.22}$$

It is, in fact, this cover that complicates the problem considerably. If it were absent, $Rd = 0$, the solution could be obtained by the mirror principle [101]:

$$G^{(Rd=0)}(\boldsymbol{r}, \boldsymbol{r}')(\boldsymbol{r}, \boldsymbol{r}') = \frac{1}{4\pi|\boldsymbol{r} - \boldsymbol{r}'|} - \frac{R_E}{4\pi r'|\boldsymbol{r} - (R_E^2/r'^2)\boldsymbol{r}'|}. \tag{4.23}$$

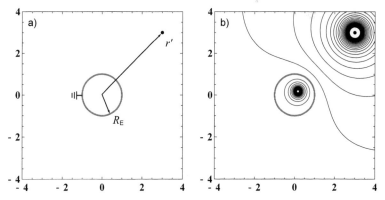

Figure 4.3: (a) Example of a unit charge at the point $\boldsymbol{r}'$ near a grounded conducting unit sphere and (b) its equipotential lines.

An example is illustrated in Figure 4.3 (a): a point charge ($\boldsymbol{r}' = 3\boldsymbol{x} + 3\boldsymbol{y}$) locates near a grounded unit sphere, which causes a mirror charge inside the sphere. The solution for the potential at an arbitrary position $\boldsymbol{r}$ is contributed from both the point charge and the mirror charge as explained by the equation (4.23). The equipotential lines are displayed in Figure 4.3 (b).

Here, however, a closed solution is not possible, and we aim instead for a series solution. Obviously, because of symmetry, an expansion into spherical harmonics $Y_{lm}(\theta, \varphi)$ is possible, so we make the ansatz (4.18), where the $R(r, r')$ is a yet unknown function.

The completeness relation is used to describe the delta function as

$$\delta^{(3)}(\boldsymbol{r} - \boldsymbol{r}') = \frac{1}{r^2}\delta(r - r')\sum_{l=0}^{\infty}\sum_{m=-l}^{l} Y_{lm}^{\star}(\theta', \varphi')Y_{lm}(\theta, \varphi). \tag{4.24}$$

Inserting (4.18) and (4.24) into (4.1) leads to

$$\frac{\partial}{\partial r}r^2\varepsilon_{\mathrm{r}}(r)\frac{\partial R_l(r, r')}{\partial r} - \varepsilon_{\mathrm{r}}(r)l(l+1)R_l(r, r') = -\delta(r - r'). \tag{4.25}$$

In intervals which contain neither R nor r', the solutions are of the form

$$R_l(r, r') = A(r')r^l + B(r')\frac{1}{r^{l+1}}. \tag{4.26}$$

According to the geometry of the IMRP, the regional radial Green's function is defined. $R_l^{(d,d)}$ and $R_l^{(p,p)}$ describe the situation that $\boldsymbol{r}$ and $\boldsymbol{r}'$ both locate either in the dielectric ($r' < R$) or plasma bulk ($r, r' > R$), whereas $R_l^{(d,p)}$ indicates that one of the vectors is in the dielectric and the other is in the plasma bulk, i.e., $r < R < r'$ or $r' < R < r$

$$R_l(r, r') = \begin{cases} R_l^{(d,d)}(r, r'), & r, r' \leq R \\[2mm] R_l^{(d,p)}(r, r'), & r < R \leq r' \text{ or } r' < R \leq r \\[2mm] R_l^{(p,p)}(r, r'), & r, r' > R \end{cases} \tag{4.27}$$

Firstly, $R_l^{(d,p)}(r, r')$ is to be discussed. Here, the general form (4.26) for $r \neq r'$ can be written as

$$R_l(r, r') = \begin{cases} A(r')r^l + \dfrac{B(r')}{r^{(l+1)}}, & R_{\mathrm{E}} \leq r < R \leq r' < \infty \\[4mm] A'(r')r^l + \dfrac{B'(r')}{r^{(l+1)}}, & R_{\mathrm{E}} \leq r' < R \leq r < \infty \end{cases} \tag{4.28}$$

The coefficients $A(r')$ and $B(r')$ are rearranged due to the vanishing of $R_l^{(d,p)}(r,r')$ at $r = R_E$, and the boundary condition at infinity suggests that $A'(r')$ goes to zero,

$$R_l^{(d,p)}(r,r') = \begin{cases} A(r')\left(r^l - \dfrac{R_E^{2l+1}}{r^{l+1}}\right), & R_E \leq r < R \leq r' < \infty \\[3mm] B'(r')\dfrac{1}{r^{l+1}}, & R_E \leq r' < R \leq r < \infty \end{cases} \tag{4.29}$$

The symmetry of $R_l^{(d,p)}(r,r')$ in r and r' requires the coefficients $A(r')$ and $B'(r')$ be such that $R_l^{(d,p)}$ can be written

$$R_l^{(d,p)}(r,r') = g_l\left(\frac{r_<^l}{r_>^{l+1}} - \frac{R_E^{2l+1}}{r^{l+1}r'^{l+1}}\right), \tag{4.30}$$

where $r_<$ ($r_>$) represents the smaller (larger) of r and r'. The effect of the delta function is considered according to equation (4.25) to determine the constant g_l, where a discontinuity exists at $r = r' = R$. It is integrated over the interval from $r = r' - \epsilon$ to $r = r' + \epsilon$, where ϵ is assumed to be a small number,

$$\varepsilon_r(\boldsymbol{r})\frac{\mathrm{d}}{\mathrm{d}r}[R_l^{(d,p)}(r,r')]\Big|_{r'-\epsilon}^{r'+\epsilon} = -\frac{1}{r^2}. \tag{4.31}$$

Taking different permittivities into account

$$g_l\left(-(l+1)\frac{r'^l}{r^{l+2}} + (l+1)\frac{R_E^{2l+1}}{r^{l+2}r'^{l+1}}\right) - g_l\varepsilon_r\left(l\frac{r'^{l-1}}{r^{l+1}} + (l+1)\frac{R_E^{2l+1}}{r^{l+2}r'^{l+1}}\right) = -\frac{1}{r^2}. \tag{4.32}$$

Therefore, we obtain

$$g_l = \frac{R^{2l+1}}{(l+1+l\varepsilon_r)R^{2l+1} + (l+1)(\varepsilon_r - 1)R_E^{2l+1}}. \tag{4.33}$$

Secondly, for $r, r' > R$, the radial functions $R_l^{(p,p)}$ are discussed in the following. Similarly, the boundary condition at infinity applies in the general form 4.26, besides, the jump conditions at $r = R$ between $R_l^{(d,p)}$ and $R_l^{(p,p)}$ are used to determine the coefficients of the radial functions $R_l^{(p,p)}$, which are written as

$$R_l^{(p,p)}\Big|_{r\to\infty} = 0, \tag{4.34}$$

and

$$\varepsilon_r\frac{\partial R_l^{(d,p)}}{\partial r}\Big|_{r=R} = \frac{\partial R_l^{(p,p)}}{\partial r}\Big|_{r=R}. \tag{4.35}$$

Then, the discontinuity at $r = r'$, and the symmetry at r and r' lead to a simplified expression of $R_l^{(p,p)}$:

$$R_l^{(p,p)}(r,r') = \frac{1}{2l+1}\left(\frac{r_<^l}{r_>^{l+1}} - f_l\frac{R^{2l+1}}{r^{l+1}r'^{l+1}}\right), \tag{4.36}$$

where

$$f_l = \frac{l(\varepsilon_r - 1)R^{2l+1} + (l + \varepsilon_r + l\varepsilon_r)R_E^{2l+1}}{(l+1+l\varepsilon_r)R^{2l+1} + (l+1)(\varepsilon_r - 1)R_E^{2l+1}}. \tag{4.37}$$

Lastly, for $r, r' < R$, the radial function $R_l^{(d,d)}$ can be obtained in the similar way. Notably, the free particles are assumed only in the simulation domain $r > R$. The boundary condition at the surface of electrodes $r = R_E$ is given, and the jump condition at $r = R$ needs to be considered

$$R_l^{(d,d)}\bigg|_{r=R_E} = 0, \tag{4.38}$$

and

$$\frac{\partial R_l^{(d,d)}}{\partial r}\bigg|_{r=R} = \frac{\partial R_l^{(d,p)}}{\partial r}\bigg|_{r=R}. \tag{4.39}$$

Contributed by the discontinuity at $r = r'$ and the symmetry of r and r', the explicit form of $R_l^{(d,d)}$ can be determined:

$$R_l^{(d,d)}(r,r') = \frac{1}{(2l+1)\varepsilon_r}\left(h_l\left(\frac{r_<^l}{r_>^{l+1}} - \frac{R_E^{2l+1}}{r^{l+1}r'^{l+1}}\right) + (h_l - 1)\left(\frac{r_<^l}{r_>^{l+1}} - \frac{r^l r'^l}{R_E^{2l+1}}\right)\right) \tag{4.40}$$

with

$$h_l = \frac{(l+1+l\varepsilon_r)R^{2l+1}}{(l+1+l\varepsilon_r)R^{2l+1} + (l+1)(\varepsilon_r - 1)R_E^{2l+1}}. \tag{4.41}$$

Consequently, the radial function $R_l(r,r')$ of IMRP can be calculated explicitly, the configuration-dependent coefficients h_l, g_l and f_l in the radial functions are defined for a compact notation.

4.3 Spectral Kinetic Model

The Green's function of the IMRP-plasma system ($R_E \leq |\boldsymbol{r}| < \infty$) is obtained explicitly. Then the basic idea is to truncate the interactions with a short distance $|\boldsymbol{r} - \boldsymbol{r}'|$ so that the coefficients in spherical harmonics can be determined. In Figure 4.4, the azimuthal symmetry denoted by azimuthal angle φ in the potential indicates the coefficient $m = 0$. Besides, the antisymmetry with respect to $\theta \to \pi - \theta$ leads to the vanishing of the expansion with the even coefficient l. Therefore, depending on the required mode number, the coefficient l can be assigned, where the projection operator $\hat{P}$ is determined as the truncation of the infinite series.

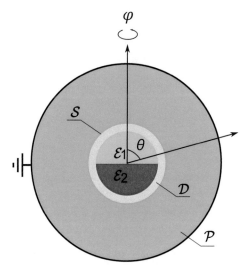

Figure 4.4: The IMRP has an azimuthal symmetry denoted by φ, and antisymmetry in respect of $\theta \to \pi - \theta$. Then the coefficients l and m in the expansion can be determined.

In the last section, the radial functions $R_l(r, r')$ in the expression (4.27) are calculated as

$$
\begin{cases}
R_l^{(d,d)}(r, r') &= \dfrac{1}{(2l+1)\epsilon_r}\left(h_l \left(\dfrac{r_<^l}{r_>^{l+1}} - \dfrac{R_E^{2l+1}}{r^{l+1}r'^{l+1}} \right) + (h_l - 1)\left(\dfrac{r_<^l}{r_>^{l+1}} - \dfrac{r^l r'^l}{R_E^{2l+1}} \right) \right) \\[2ex]
R_l^{(d,p)}(r, r') &= g_l \left(\dfrac{r_<^l}{r_>^{l+1}} - \dfrac{R_E^{2l+1}}{r^{l+1}r'^{l+1}} \right) \\[2ex]
R_l^{(p,p)}(r, r') &= \dfrac{1}{2l+1}\left(\dfrac{r_<^l}{r_>^{l+1}} - f_l \dfrac{R^{2l+1}}{r^{l+1}r'^{l+1}} \right)
\end{cases}
\qquad , \quad (4.42)
$$

with the coefficients

$$
\begin{cases}
h_l = \dfrac{(l+1+l\varepsilon_{\mathrm{r}})R^{2l+1}}{(l+1+l\varepsilon_{\mathrm{r}})R^{2l+1} + (l+1)(\varepsilon_{\mathrm{r}}-1)R_{\mathrm{E}}^{2l+1}} \\[4mm]
g_l = \dfrac{R^{2l+1}}{(l+1+l\varepsilon_{\mathrm{r}})R^{2l+1} + (l+1)(\varepsilon_{\mathrm{r}}-1)R_{\mathrm{E}}^{2l+1}} \\[4mm]
f_l = \dfrac{l(\varepsilon_{\mathrm{r}}-1)R^{2l+1} + (l+\varepsilon_{\mathrm{r}}+l\varepsilon_{\mathrm{r}})R_{\mathrm{E}}^{2l+1}}{(l+1+l\varepsilon_{\mathrm{r}})R^{2l+1} + (l+1)(\varepsilon_{\mathrm{r}}-1)R_{\mathrm{E}}^{2l+1}}
\end{cases} \qquad . \tag{4.43}
$$

Hence, the potential in the field solver can be obtained with the Green's function in the form of an expansion of the spherical harmonics (4.18). However, to simulate one accurate situation with the consideration of a large number of particles, an optimized field solver is needed.

Considering the symmetry in the IMRP model, the symmetric part $\bar{\Psi}(r)$ of the characteristic function is defined as

$$
\bar{\Psi}(r) = \Psi_1(\boldsymbol{r}) + \Psi_2(\boldsymbol{r}), \tag{4.44}
$$

and the antisymmetric part $\Delta\Psi(\boldsymbol{r})$ is

$$
\Delta\Psi(\boldsymbol{r}) = \frac{1}{2}\Big(\Psi_1(\boldsymbol{r}) - \Psi_2(\boldsymbol{r})\Big) \tag{4.45}
$$

Similarly, the applied voltages at both electrodes of the IMRP are separated into a symmetric part $\bar{u}(t)$,

$$
\bar{u}(t) = \frac{1}{2}\Big((u_1(t) + u_2(t)\Big), \tag{4.46}
$$

and an asymmetric part $\Delta u(t)$

$$
\Delta u(t) = u_1(t) - u_2(t). \tag{4.47}
$$

Inserting (4.44) and (4.46) into (4.13), the potential can be arranged as following:

$$
\Phi(\boldsymbol{r}, t) = \frac{1}{\varepsilon_0} \sum_{i=1}^{N} q_i \hat{\mathrm{P}} G(\boldsymbol{r}, \boldsymbol{r}_i(t)) + \Phi_{\mathcal{S}}(r) + \bar{u}(t)\bar{\Psi}(r) + \Delta u(t)\hat{\mathrm{P}}\Delta\Psi(\boldsymbol{r}). \tag{4.48}
$$

Here, the corresponding projection operator $\hat{\mathrm{P}}$ can be eliminated since the potential of surface charge $\Phi_{\mathcal{S}}(r)$ and the symmetric characteristic function $\bar{\Psi}(r)$ are defined in the static condition ($l = 0$ and $m = 0$). In this section, the explicit expression of $\bar{\Psi}(r)$, $\hat{\mathrm{P}}\Delta\Psi(\boldsymbol{r})$, $\bar{u}(t)$, and $\Delta u(t)$, in addition to the potential of surface charge $\Phi_{\mathcal{S}}(r)$, are derived to obtain a mathematically compact expression for the potential.

4.3.1 Potential of Surface Charge

Starting with the potential of surface charge $\Phi_\mathcal{S}(r)$, it is established in the form of Green's function, which can be written as:

$$\Phi_\mathcal{S}(r) = \frac{1}{\varepsilon_0} \int_\mathcal{S} \sigma_\mathcal{S}\, G(\boldsymbol{r}, \boldsymbol{r}')\, \mathrm{d}f' = \frac{Q_\mathcal{S}}{4\pi\varepsilon_0 R^2} \int_\mathcal{S} G(\boldsymbol{r}, \boldsymbol{r}')\, \mathrm{d}f'. \tag{4.49}$$

For the case $r < R$, according to the assumption, $r' \rightarrow R$ is applied, then $R_l^{(d,p)}(r, r')$ can be used to determine the solution:

$$\Phi_\mathcal{S}(r)|_{r<R} = \frac{Q_\mathcal{S}}{4\pi\varepsilon_0} R_{l=0}^{(d,p)}(r, r') \tag{4.50}$$

$$= \frac{Q_\mathcal{S}}{4\pi\varepsilon_0} \left(\frac{1}{R(d + \varepsilon_\mathrm{r} - d\varepsilon_\mathrm{r})} - \frac{1-d}{r(d + \varepsilon_\mathrm{r} - d\varepsilon_\mathrm{r})} \right).$$

Similarly, for the case $r > R$, the Greens function is in the form of $R_{l=0}^{(p,p)}(r, r')$, we obtain

$$\Phi_\mathcal{S}(r)|_{r>R} = \frac{Q_\mathcal{S}}{4\pi\varepsilon_0} R_{l=0}^{(p,p)}(r, r') \tag{4.51}$$

$$= \frac{Q_\mathcal{S}}{4\pi\varepsilon_0} \left(\frac{d}{r(d + R\varepsilon_\mathrm{r} - d\varepsilon_\mathrm{r})} \right).$$

4.3.2 Characteristic Functions

Then the characteristic functions $\Psi(\boldsymbol{r})$ is discussed to provide the expression of $\bar{\Psi}(r)$ and $\hat{P}\Delta\Psi(\boldsymbol{r})$. To be more specific, according to the definition, the characteristic functions describe the geometry of the IMRP, which obey the Laplace equation in domain $\mathcal{V}$. It indicates that $\Psi(\boldsymbol{r})$ can be expanded in spherical harmonics as

$$\Psi(\boldsymbol{r}) = \sum_{l=0}^{\infty} \sum_{m=-l}^{l} \psi_l(r) Y_{lm}^\star(\theta', \varphi') Y_{lm}(\theta, \varphi), \tag{4.52}$$

where the general solutions in r-direction are

$$\psi_l(r) = \begin{cases} a_l^\mathcal{D} \dfrac{r^l}{R^l} + b_l^\mathcal{D} \dfrac{R^{l+1}}{r^{l+1}}, & R_\mathrm{E} \leq r < R \\[2ex] a_l^\mathrm{vac} \dfrac{r^l}{R^l} + b_l^\mathrm{vac} \dfrac{R^{l+1}}{r^{l+1}}, & R \leq r < +\infty \end{cases}. \tag{4.53}$$

The defined coefficients $a_l^\mathcal{D}$, $b_l^\mathcal{D}$ relate to the situation in the dielectric whereas a_l^vac $b_l^\mathcal{D}$ describe the vacuum condition. The continuity of the vacuum potential and the electric flux density at the surface of the dielectric suggest the following transition conditions

$$\psi_l^\mathcal{D}(R) = \psi_l^\mathrm{vac}(R), \tag{4.54}$$

and

$$\left.\frac{\partial \psi_l^{\mathcal{D}}}{\partial r}\right|_{r=R} = \left.\frac{\partial \psi_l^{\mathrm{vac}}}{\partial r}\right|_{r=R}. \tag{4.55}$$

Besides, according to the boundary conditions at infinity and the electrodes (4.11), the coefficients can be evaluated as

$$\begin{cases} a_l^{\mathcal{D}} = \dfrac{(l+1)(\varepsilon_r - 1)(1-d)^{(l+1)}}{(l+1)(\varepsilon_r - 1)(1-d)^{(2l+1)} + l\varepsilon_r + l + 1} &, \\[4mm] b_l^{\mathcal{D}} = \dfrac{(1-d)^{(l+1)}(l\varepsilon_r + l + 1)}{(l+1)(\varepsilon_r - 1)(1-d)^{(2l+1)} + l\varepsilon_r + l + 1} &, \\[4mm] a_l^{\mathrm{vac}} = 0 &, \\[4mm] b_l^{\mathrm{vac}} = \dfrac{(2l+1)\varepsilon_r (1-d)^{(l+1)}}{(l+1)(\varepsilon_r - 1)(1-d)^{(2l+1)} + l\varepsilon_r + l + 1} &. \end{cases} \tag{4.56}$$

Obviously, symmetric characteristic function $\bar{\Psi}(r)$ can be directly calculated from the expression of $\Psi(\boldsymbol{r})$, since it represents the static situation ($l = 0$). The explicit form is

$$\bar{\Psi}(r) = \begin{cases} \dfrac{1-d}{d + \epsilon_r - d\epsilon_r}\dfrac{R}{r} + \dfrac{(1-d)(\epsilon_r - 1)}{d + \epsilon_r - d\epsilon_r}, & R_{\mathrm{E}} \leq r < R \\[4mm] \dfrac{(1-d)\epsilon_r}{d + \epsilon_r - d\epsilon_r}\dfrac{R}{r}, & R \leq r < +\infty \end{cases}. \tag{4.57}$$

The antisymmetric function $\Delta\Psi(\boldsymbol{r})$ is expressed in the form of Legendre series $P_l(x)$. The coefficient c_l is determined by means of Rodrigues' formula [101] which contains the information about the electrode configuration within the probe tip. The antisymmetric excitation at the electrodes indicates only the odd terms are different from zero, we obtain

$$\Delta\Psi(\boldsymbol{r}) = \sum_l c_l \psi_l(r) P_l(\cos\theta) \tag{4.58}$$

with the coefficient

$$c_l = \left(-\frac{1}{2}\right)^{\frac{l+3}{2}} \frac{(2l+1)(l-2)!!}{(l/2 + 1/2)!}.$$

4.3.3 Applied Voltage

The signal provided by the probe is denoted by a symmetric part $\bar{u}(t)$ and an asymmetric part $\Delta u(t)$. Specifically, $\bar{u}(t)$ is the floating potential, and $\Delta u(t)$ plays the role as the input of the simulation. It is helpful to define and analyze the charge on the electrodes Q_n as the response of the system, where $\bar{Q}$ represents the total charge on the electrodes as

$$\bar{Q}(t) = Q_1(t) + Q_2(t) \tag{4.59}$$

and ΔQ is the charge difference

$$\Delta Q(t) = \frac{1}{2}\Big(Q_1(t) - Q_2(t)\Big) \tag{4.60}$$

When the probe is in the plasma without excitation, the floating potential $\bar{u}(t)$ can be determined from $\bar{Q}(t)$ in the static condition. After the probe is switched on, the perturbation occurs, which is captured in $\Delta Q(t)$.

According to Gauss's law, the charge on the electrode $Q_n(t)$ can be calculated

$$Q_n = \int_{\mathcal{E}_n} \varepsilon_0 \varepsilon_\mathrm{r}(\boldsymbol{r}) \boldsymbol{E} \cdot \mathrm{d}\boldsymbol{f}, \tag{4.61}$$

which gives

$$Q_n = -\int_{\mathcal{E}_n} \varepsilon_0 \varepsilon_\mathrm{r}(\boldsymbol{r}) \nabla \Phi(\boldsymbol{r}) \cdot \mathrm{d}\boldsymbol{f}. \tag{4.62}$$

Inserting the expression of potential (4.9) into (4.62), we obtain

$$Q_n(t) = -\int_{\mathcal{E}_n} \varepsilon_\mathrm{r}(\boldsymbol{r}) \nabla \left(\sum_{i=1}^{N} q_i\, G(\boldsymbol{r}, \boldsymbol{r}_i(t)) + \int_{\mathcal{S}} \sigma_{\mathcal{S}}\, G(\boldsymbol{r}, \boldsymbol{r}')\, \mathrm{d}f' \right.$$
$$\left. + \int_{\mathcal{E}} \Phi(\boldsymbol{r}') \varepsilon_0 \varepsilon_\mathrm{r}(\boldsymbol{r}') \nabla' G(\boldsymbol{r}, \boldsymbol{r}') \cdot \mathrm{d}\boldsymbol{f}' \right) \cdot \mathrm{d}\boldsymbol{f}. \tag{4.63}$$

The expression of $Q_n(t)$ consists of three parts:

$$Q_n^{(1)}(t) = -\int_{\mathcal{E}_n} \varepsilon_\mathrm{r}(\boldsymbol{r}) \nabla \left(\sum_{i=1}^{N} q_i\, G(\boldsymbol{r}, \boldsymbol{r}_i(t)) \cdot \mathrm{d}\boldsymbol{f}' \right) \cdot \mathrm{d}\boldsymbol{f}, \tag{4.64}$$

$$Q_n^{(2)}(t) = -\int_{\mathcal{E}_n} \varepsilon_\mathrm{r}(\boldsymbol{r}) \nabla \left(\int_{\mathcal{S}} \sigma_{\mathcal{S}}\, G(\boldsymbol{r}, \boldsymbol{r}')\, \mathrm{d}f' \right) \cdot \mathrm{d}\boldsymbol{f}, \tag{4.65}$$

$$Q_n^{(3)}(t) = -\int_{\mathcal{E}_n} \varepsilon_\mathrm{r}(\boldsymbol{r}) \nabla \left(\int_{\mathcal{E}} \Phi(\boldsymbol{r}') \varepsilon_0 \varepsilon_\mathrm{r}(\boldsymbol{r}') \nabla' G(\boldsymbol{r}, \boldsymbol{r}') \cdot \mathrm{d}\boldsymbol{f}' \right) \cdot \mathrm{d}\boldsymbol{f}. \tag{4.66}$$

Taking the definition of the characteristic function (4.10), the first term in $Q_n(t)$ is

$$Q_n^{(1)}(t) = -\sum_{i=1}^{N} q_i \Psi(\boldsymbol{r}_i(t)) \tag{4.67}$$

which describes the influence of the free particles on the charge on the electrodes.

The effect of surface charge on the electrode n is written as $Q_n^{(2)}(t)$:

$$Q_n^{(2)}(t) = -\int_{\mathcal{S}} \sigma_{\mathcal{S}} \Psi_n(\boldsymbol{r}) \mathrm{df} \tag{4.68}$$

Since the surface charge $Q_{\mathcal{S}}$ is homogeneously distributed on $\mathcal{S}$, the influence of surface charge on both electrodes is defined as $Q_{\mathcal{E}}^{\mathcal{S}} = Q_1^{(2)} + Q_2^{(2)}$, which can be determined from the symmetric characteristic function

$$\begin{aligned}
Q_{\mathcal{E}}^{\mathcal{S}} &= -\int_{\mathcal{S}} \sigma_{\mathcal{S}} \left(\Psi_1(\boldsymbol{r}) + \Psi_2(\boldsymbol{r})\right) \mathrm{df} \\
&= -\int_{\mathcal{S}} \sigma_{\mathcal{S}} \bar{\Psi}(r) \mathrm{df}.
\end{aligned} \tag{4.69}$$

Here, the floating potential is to be calculated, which is in the plasma bulk, therefore, the the case $R \leq r < +\infty$ is considered,

$$Q_{\mathcal{E}}^{\mathcal{S}} = -Q_{\mathcal{S}} \frac{(1-d)\varepsilon_{\mathrm{r}}}{d + \varepsilon_{\mathrm{r}} - d\varepsilon_{\mathrm{r}}}. \tag{4.70}$$

Due to the homogeneity of the surface charge, $Q_n^{(2)}(t)$ can be simplified as

$$Q_n^{(2)}(t) = \frac{1}{2} Q_{\mathcal{E}}^{\mathcal{S}}. \tag{4.71}$$

The voltage from the probe denotes as $Q_n^{(3)}(t)$: the capacity coefficient $C_{nn'}$ can be written in the following

$$\begin{aligned}
C_{nn'} &= -\varepsilon_0 \int_{\mathcal{E}_n} \int_{\mathcal{E}_n'} \varepsilon_{\mathrm{r}}(\boldsymbol{r}) \varepsilon_{\mathrm{r}}(\boldsymbol{r}') \nabla \nabla' G(\boldsymbol{r}, \boldsymbol{r}') \cdot \mathrm{d}\boldsymbol{f}' \cdot \mathrm{d}\boldsymbol{f} \\
&= -\int_{\mathcal{E}_n'} \left(\varepsilon_0 \varepsilon_{\mathrm{r}}(\boldsymbol{r}) \nabla \Psi_n(\boldsymbol{r})\right) \cdot \mathrm{d}\boldsymbol{f}' \\
&= -\int_{\mathcal{V}} \left(\varepsilon_0 \varepsilon_{\mathrm{r}}(\boldsymbol{r}) \nabla \Psi_n(\boldsymbol{r}) \cdot \nabla \Psi_{n'}(\boldsymbol{r})\right) \mathrm{d}^3 r.
\end{aligned} \tag{4.72}$$

which is proven symmetric.

Then, the term describes the signal from the probe can be obtained as

$$Q_n^{(3)}(t) = \sum_{n'=1}^{2} C_{nn'} u_n'(t), \tag{4.73}$$

Thus, the expression of the charge on the electrodes can be simplified by replacing the corresponding characteristic function as

$$Q_n(t) = -\sum_{i=1}^{N} q_i \Psi(\mathbf{r}_i(t)) + \frac{1}{2}Q_{\mathcal{E}}^{\mathcal{S}} + \sum_{n'=1}^{2} C_{nn'}u'_n(t), \qquad (4.74)$$

To be more specific, the charge on each electrode of the IMRP is

$$Q_1 = -\sum_{i=1}^{N} q_i \Psi_1(\mathbf{r}_i(t)) + \frac{1}{2}Q_{\mathcal{E}}^{\mathcal{S}} + (C_{11} + C_{12})\bar{u}(t) + (C_{11} - C_{12})\frac{\Delta u(t)}{2}, \qquad (4.75)$$

and

$$Q_2 = -\sum_{i=1}^{N} q_i \Psi_2(\mathbf{r}_i(t)) + \frac{1}{2}Q_{\mathcal{E}}^{\mathcal{S}} + (C_{21} + C_{22})\bar{u}(t) + (C_{21} - C_{22})\frac{\Delta u(t)}{2}. \qquad (4.76)$$

It is convenient to define the total capacitance as

$$\bar{C} = C_{11} + C_{12} + C_{21} + C_{22}, \qquad (4.77)$$

and the capacitance in vacuum

$$C_{\text{vac}} = \frac{C_{11} - C_{12}}{2} = \frac{C_{22} - C_{21}}{2}. \qquad (4.78)$$

The self capacitance of a conducting sphere can be calculated as $\bar{C}$, and C_{vac} is canceled out due to the symmetry of the capacity coefficients. Then the floating potential u_{sym} can be calculated by assuming the total charge on the electrodes in static situation as zero,

$$\bar{Q} = Q_1 + Q_2 = 0. \qquad (4.79)$$

The explicit expression is

$$\bar{Q} = -\frac{(1-d)R\varepsilon_r}{d + \varepsilon_r - d\varepsilon_r} \sum_{i=1}^{N} \frac{q_i}{r_i} - Q_S \frac{(1-d)\varepsilon_r}{d + \varepsilon_r - d\varepsilon_r} + \frac{4\pi\varepsilon_0 R(1-d)\varepsilon_r}{d + \varepsilon_r - d\varepsilon_r} \cdot \bar{u} = 0 \qquad (4.80)$$

As a result, the floating potential of the system is derived

$$\bar{u} = \frac{1}{4\pi\varepsilon_0} \left(\frac{Q_S}{R} + \sum_{i=1}^{N} \frac{q_i}{r_i} \right). \qquad (4.81)$$

Then the asymmetric part $\Delta u(t)$ of the signal from the IMRP can be set accordingly in the simulation.

4.3.4 Modified Spectral Kinetic Model

Finally, the expression of the potential can be simplified by summing up the calculated expressions. Inserting the expression of $\bar{u}$ (4.81) into (4.48), the potential in the domain $\mathcal{V}$ is

$$
\Phi(\boldsymbol{r}, t) = \frac{1}{\varepsilon_0} \sum_{i=1}^{N} q_i \hat{P} G(\boldsymbol{r}, \boldsymbol{r}_i(t)) + \Phi_{\mathcal{S}}(r)
$$
$$
+ \frac{1}{4\pi\varepsilon_0} \left(\frac{Q_{\mathcal{S}}}{R} + \sum_{i=1}^{N} \frac{q_i}{r_i} \right) \bar{\Psi}(r) + \Delta u(t) \hat{P} \Delta \Psi(\boldsymbol{r}). \tag{4.82}
$$

As can be seen, the modified potential can be completed with the influence of the surface charge $\Phi_{\mathcal{S}}(r)$ (4.50), (4.51), the floating potential $\bar{u}$ (4.81), and the symmetric characteristic function (4.57). The following expression can be obtained

$$
\Phi_{\mathcal{S}}(r) + \frac{1}{4\pi\varepsilon_0} \frac{Q_{\mathcal{S}}}{R} \bar{\Psi}(r) = \begin{cases} \dfrac{Q_{\mathcal{S}}}{4\pi\varepsilon_0 R}, & R_{\mathrm{E}} \leq r < R \\[3mm] \dfrac{Q_{\mathcal{S}}}{4\pi\varepsilon_0 r}, & R \leq r < +\infty \end{cases}, \tag{4.83}
$$

and a new modified Green's function can be defined as

$$
\tilde{G}(\boldsymbol{r}, \boldsymbol{r}') = \hat{P} G(\boldsymbol{r}, \boldsymbol{r}') + \frac{1}{4\pi r'} \bar{\Psi}(r). \tag{4.84}
$$

Focusing on the interaction within the plasma, it is reasonable to only investigate the case $R \leq r < +\infty$ of the potential

$$
\Phi(\boldsymbol{r}, t) = \frac{1}{\varepsilon_0} \sum_{i=1}^{N} q_i \tilde{G}(\boldsymbol{r}, \boldsymbol{r}_i(t)) + \frac{Q_{\mathcal{S}}}{4\pi\varepsilon_0 r} + \Delta u(t) \hat{P} \Delta \Psi(\boldsymbol{r}). \tag{4.85}
$$

According to the results in [17, 47], the prominent feature of the resonance spectrum of the IMRP is the absorption peaks. The first absorption peak, the so-called dipole mode, is the dominant one. The resonances of higher modes are barely visible. Therefore, in addition to the static situation ($l = 0$), considering only the dipole mode ($l = 1$) in the simulation can be seen as an applicable approximation. Thus, the projection operator $\hat{P}$ is determined, which leads to a simplified model with the truncated Green's function

$$
\tilde{G}(\boldsymbol{r}, \boldsymbol{r}') = \frac{1}{4\pi} \left(\frac{1}{r_>} + \frac{1}{4\pi} \left(\frac{r_<}{r_>^2} - f_1 \frac{R^3}{r^2 r'^2} \right) \cos\theta \cos\theta' \right). \tag{4.86}
$$

Therefore, after inserting the investigated potential (4.85) with the truncated Green's function (4.86) into the equation (4.17), the modified equations of motion for particle k are obtained as

$$\frac{\mathrm{d}\boldsymbol{r}_k}{\mathrm{d}t} = \boldsymbol{v}_k, \tag{4.87}$$

$$m_k \frac{\mathrm{d}\boldsymbol{v}_k}{\mathrm{d}t} = -\frac{1}{\varepsilon_0} \sum_{\substack{i=1 \\ i \neq k}}^{N} q_k q_i \, \nabla_k \tilde{G}(\boldsymbol{r}_k, \boldsymbol{r}_i) - q_k \nabla_k \frac{Q_S}{4\pi\varepsilon_0 r_k} - q_k \Delta u(t) \nabla_k \Delta\Psi(\boldsymbol{r}_k).$$

Apparently, the scheme is highly optimized due to the symmetry of the IMRP model, where the symmetry part of the voltage $\bar{u}(t)$ and the symmetric characteristic function $\bar{\Psi}(r)$ are omitted in the modified equations of motion. Thus, the asymmetric part of the voltage $\Delta u(t)$ can be defined as the input signal in the simulation, and the charge difference is captured as the response of the input, which writes

$$\Delta Q(t) = -\sum_{i=1}^{N} q_i \Delta\Psi(\boldsymbol{r}_i) + C_{\mathrm{vac}} \Delta u(t). \tag{4.88}$$

Notably, the spectral response of the system is expected to demonstrate the influence of kinetic effects. That is, the output of the simulation is required to be in the frequency domain. However, the calculations in the frequency domain are time-consuming regarding the convergence of a sequence of periodic functions. Therefore, the charge difference on the electrons $\Delta Q(t)$ in the spectral kinetic simulation is recorded in the time domain. Then, the numerical results are converted from the time domain into the frequency domain. In this approach, the complete frequency can be covered in one simulation. To be more precise, the efficient strategy in the spectral kinetic scheme is proposed: an impulse signal is provided as the input of the system, then the charge on the electrodes in the time domain is calculated as the output. By applying a Fourier transform, the impulse response of the IMRP-plasma system can be expressed in the frequency domain for further analysis.

5 Numerical Simulation

In this chapter, the necessary settings for the implementation of the proposed spectral kinetic simulation are discussed. As stated, the equations used to describe the IMRP-plasma system can be solved analytically, or they can be solved numerically. However, the issue arises in attempting to obtain the exact solutions due to the complexity of the model. That is, solving the integration of the Vlasov equation is cumbersome in terms of calculations. Therefore, particle simulation using numerical methods is preferred in this work. By appropriately taking the initial numerical assumption, the efficiency of the simulation is optimized to a substantial extent.

Basically, the idea of the spectral kinetic simulation is to record the response of the probe-plasma system after an impulse is provided via IMRP. Then the analysis of the simulated resonance structure of plasma is expected to explain the physical phenomena observed in the experiments. The input is assumed to be small so that the linear behavior of the plasma is captured. The system allows converting the output from the time domain into the frequency domain for the direct evaluation of the plasma parameters. This is often known as the linear time-invariant system, which is introduced shortly.

Besides, the details of initial conditions in the kinetic simulation are clarified in Section 5.2, including the distribution of the particles in the coordinate and velocity space. To resolve the issue of long computational time, electron and ion density profiles of the steady-state are calculated from the equilibrium under floating conditions. Then the newly generated particles are distributed according to these numerical results. It avoids the unnecessary computation of sheath formation. Especially since the ions are comparatively slow, this approximation can drastically reduce the simulation time. Additionally, by solving the Boltzmann-Poisson equation, the static charge density and sheath thickness can be obtained from the given plasma parameters. Furthermore, the particle interactions with the surface of the MRP and the outer boundary are of great importance in such kinetic simulations. Hence, the possible strategy of boundary conditions is discussed.

Briefly speaking, the essential information to describe the static status of the system is determined. Particularly, the corresponding initialized particle densities in the simulation are monitored. With the assumption of the charge equilibrium, the explicit initialization of the simulation is demonstrated: The complete simulation domain with the ions and electrons in the form of super-particles around an idealized MRP is visualized.

5.1 Linear Time-Invariant System

In the study of the spectral kinetic simulation of IMRP, ideally, an impulse can be given as the signal provided by the probe to the system in the form of the Dirac delta function (δ−function), then the impulse response of the system can be obtained. Apparently, the linear behavior of the plasma-probe system is of particular importance. As shown in Figure 5.1, the concept of the linear time-invariant system can be applied: The output is linear to the input, and its value is not relevant to the time that input is applied. In the simulation, the IMRP performs a "kick" to all the particles by applying an impulse, which leads to a resonance behavior of the plasma. The output at any time only depends on the past and present values of input in such a causal system. That is, any delay in the input is reflected in the output. In the spectral kinetic simulation, the charge on the electrodes is directly recorded in the time domain as the output of the system. However, as is introduced, the resonance curve in the frequency domain is to be investigated since the corresponding plasma parameters are directly related to the resonance frequency and the full width at half maximum. Fortunately, the continuous-time system can also be characterized in the frequency domain by the transfer function, which can be either Laplace transform or Fourier transform. In this section, the causal linear time-invariant system is introduced to describe the IMRP-plasma system, and the relation between the time domain and frequency domain is discussed.

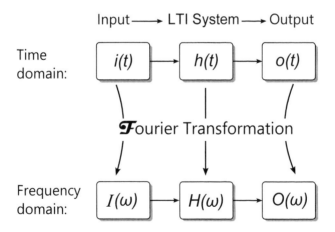

Figure 5.1: Block diagram of a causal linear time-invariant system, relation between the time domain and frequency domain is presented.

Assuming that a linear time-invariant system in the time domain is with input signal $i(t)$ and output signal $o(t)$, the behavior of such a system can be described by the convolution integral [102]:

$$o(t) = \int_{-\infty}^{t} h(t - t')i(t')dt'. \tag{5.1}$$

Substituting the term $t - t' = \tau$, the causality suggests $h(\tau)$ equals to zero for $\tau < 0$. The convolution integral can be written as

$$o(t) = \int_{-\infty}^{+\infty} h(\tau)i(t - \tau)d\tau, \tag{5.2}$$

where $h(t)$ is the response of the system to the impulse.

According to the convolution theorem, the output of the system in the frequency domain $O(\omega)$ is obtained by the Fourier transform

$$
\begin{aligned}
\mathcal{F}\{(h(\tau) * i(t - \tau))(t)\} &= \int_{-\infty}^{+\infty} h(\tau) * i(t - \tau)\, e^{-i\omega t}\, dt \\
&= \int_{-\infty}^{+\infty} \int_{-\infty}^{+\infty} h(\tau)i(t - \tau)\, d\tau\, e^{-i\omega t}\, dt \\
&= O(\omega).
\end{aligned}
\tag{5.3}
$$

The impulse and response of it can be expressed in the frequency domain as $I(\omega)$ and $H(\omega)$, respectively:

$$i(t - \tau) = \frac{1}{2\pi} \int_{-\infty}^{+\infty} I(\omega')e^{i\omega'(t-\tau)}\, d\omega', \tag{5.4}$$

and

$$h(\tau) = \frac{1}{2\pi} \int_{-\infty}^{+\infty} H(\omega'')e^{i\omega''\tau}\, d\omega''. \tag{5.5}$$

Inserting equation (5.4) and equation (5.5) into equation (5.3), the expression of $O(\omega)$ is expanded as

$$O(\omega) = \frac{1}{(2\pi)^2} \int_{-\infty}^{+\infty} \int_{-\infty}^{+\infty} \int_{-\infty}^{+\infty} H(\omega'')e^{i\omega''\tau} \int_{-\infty}^{+\infty} I(\omega')e^{i\omega'(t-\tau)}e^{-i\omega t}\, d\omega''\, d\omega'd\tau\, dt. \tag{5.6}$$

Here, $\delta-$function is given as

$$\delta(\omega'' - \omega') = \frac{1}{2\pi} \int_{-\infty}^{+\infty} e^{i(\omega''-\omega')\tau}d\,\tau, \tag{5.7}$$

which simplifies the function $O(\omega)$

$$O(\omega) = \frac{1}{2\pi} \int_{-\infty}^{+\infty} \int_{-\infty}^{+\infty} H(\omega')I(\omega')e^{i(\omega'-\omega)t}\, d\omega'\, dt. \tag{5.8}$$

Similarly, $\delta-$function for $\omega' - \omega$ writes

$$\delta(\omega' - \omega) = \frac{1}{2\pi} \int_{-\infty}^{+\infty} e^{i(\omega' - \omega)t} \, \mathrm{d}t, \tag{5.9}$$

which is inserted into equation (5.6).

Eventually, the functions to describe the system in the frequency domain can be obtained. The linearity of the output to the input can be proved in the expression

$$O(\omega) = H(\omega) \cdot I(\omega). \tag{5.10}$$

The relation between the time frequency domains is presented so that the spectral kinetic simulation can be implemented in the time domain while the output is investigated in the frequency domain. Thus, the response of the system, which is the charge on the electrodes, for all the frequencies can be determined in one execution.

5.2 Initial Conditions

5.2.1 Steady State of the System

Before the first execution of the cycle, it is important to specify the initial condition appropriately so that the simulation can be both efficient and accurate. It includes the initial distributions of the particles in the coordinate and velocity space. The particles in a simulation can be generated by initial loading or particle injection from a volumetric source. Regarding corresponding velocities of the particles, they are calculated from a given velocity distribution function. In this section, those distributions are discussed. Conceptually, the straightforward strategy is to randomly distribute the particles, including ions and electrons, in the spatial coordinates within the simulation volume. Then the static state of the system before the onset of the electric signal can be reached over a certain period of time, i.e., the plasma sheath is formed since most electrons are depleted from the boundary interface between the surface of electrodes and plasma due to the directional electric field. It indicates that the particles in the sheath region are mostly positive ions and neutrals. However, the obvious drawback would be the higher requirement of computational effort. Considering the fact that the ions are extremely slow compared to the electrons, the simulation time consumed for ions to form the sheath is much longer than solely to record the response of the system. As a matter of fact, the impulse and the system's response are of particular importance for the characterization of the investigated plasma; the dynamics of particles before receiving the signal are insignificant. Therefore, focusing only on the simulation of the resonance behavior is beneficial in decreasing the computational costs.

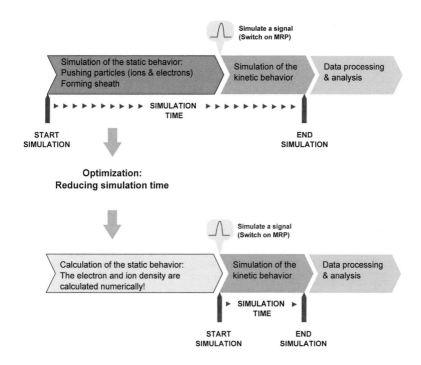

Figure 5.2: Optimization of the simulation: the ion and electron density profiles are numerically calculated as the initial condition so that the simulation time can be remarkably reduced.

Apparently, the optimization of the simulation strategy is necessary. As is shown in Figure 5.2, to reduce the simulation time, the steady-state of the IMRP-plasma system before the applied signal can be determined numerically from the Boltzmann-Poisson equation. Then the particles are distributed according to the calculated electron and ion density profile. Thus, the simulation starts from the point when the probe is switched on, i.e., the electric signal is given as a "kick" to the plasma via the IMRP.

In particular, the initial state before the onset of the electric signal corresponds to a spherically symmetric probe-plasma equilibrium under floating conditions. The Poisson equation relates the potential $\Phi(r)$ to the ion and electron charge densities:

$$-\varepsilon_0 \frac{1}{r^2} \frac{\partial}{\partial r} r^2 \frac{\partial \Phi(r)}{\partial r} = e(n_i(r) - n_e(r)). \tag{5.11}$$

The electron density is described by the Boltzmann relation. Notably, denoting the particle density far away from the probe by n_∞ and choosing the potential reference there to 0, it reads

$$n_e = n_\infty \exp\left(\frac{e\Phi}{T_e}\right). \tag{5.12}$$

Here, n_∞ is given to define the quasi neutrality, where the plasma can be treated as being electrically neutral.

To derive the expression of ion density, the ion continuity equation is considered

$$\frac{\partial n_i}{\partial t} + \nabla \cdot (n_i v_i) = 0. \tag{5.13}$$

The assumption of the stationary state indicates that n_i is a constant over time. Then the continuity equation is simplified as

$$\nabla \cdot (n_i v_i) = 0. \tag{5.14}$$

The ion flux to the probe is spatially constant as the ionization or recombination is negligible. We write it as a product of the density n_∞, the Bohm velocity $\sqrt{T_e/m_i}$, and an yet unknown constant $\hat{R}$:

$$4\pi r^2 n_i v_i = -4\pi \hat{R}^2 n_\infty \sqrt{\frac{T_e}{m_i}}. \tag{5.15}$$

The ion velocity can be derived from energy conservation

$$\frac{1}{2} m_i v_i^2 + e\Phi = 0. \tag{5.16}$$

The ion density is then obtained as

$$n_i = \frac{\hat{R}^2}{r^2} n_\infty \sqrt{-\frac{T_e}{2e\Phi}}. \tag{5.17}$$

Inserting (5.12) and (5.17) into (5.11), the Poisson equation can be expressed as

$$-\varepsilon_0 \frac{1}{r^2} \frac{\partial}{\partial r} r^2 \frac{\partial \Phi(r)}{\partial r} = e n_\infty \left(\frac{\hat{R}^2}{r^2} \sqrt{-\frac{T_e}{2e\Phi(r)}} - \exp\left(\frac{e\Phi(r)}{T_e}\right)\right). \tag{5.18}$$

To find the solution to the ordinary differential equation (ODE), the boundary conditions need to be considered. Besides the vanishing of the potential at infinity, the potential at the surface of the electrodes ($r = R$) can be determined since the net flux is zero,

$$\Phi\Big|_{r\to+\infty} = 0, \qquad n_e v_e\Big|_{r=R} = n_i v_i\Big|_{r=R}. \tag{5.19}$$

In order to calculate the electron flux, the distribution in the velocity space is required. Here, with the electron thermal velocity $v_{\mathrm{th,e}} = \sqrt{2T_e/m_e}$, the Maxwell velocity distribution function is applied, which writes

$$f_{\mathrm{Maxwell}}(\boldsymbol{v}) = \frac{n_e}{(2\pi T_e/m_e)^{3/2}} \exp\left(-\frac{1}{2}m_e v^2 / T_e\right). \tag{5.20}$$

This is a simple assumption: It follows the fact that the physics under study happens after the formation of plasma, which is the expected static condition. In actuality, the energy distribution is more complicated for specific cases; for example, in capacitively coupled plasma or inductively coupled plasma, this assumption fails due to the existence of the high energetic electrons. However, the focus of the simulation is the impulse response of the plasma-probe system as a general case. In the simulation, the equilibrium distribution is perturbed due to the excitation of the probe. Since the simulated scenario occurs on a short time scale and the input signal is very small, after the oscillation, the electrons return to Maxwellian. Conceptually, for complicated cases, this model can be modified to meet the requirements. On a practical level, Maxwell distribution can be considered as a reliable and appropriate approximation in this work.

Accordingly, the electron flux at $r = R$ is calculated with the integration over velocity space. The floating condition states that the electron flux and the ion flux to the probe are equal. Employing the Hertz-Langmuir formula [103], it is demanded that

$$4\pi \hat{R}^2 n_\infty \sqrt{\frac{T_e}{m_i}} = 4\pi R^2 \sqrt{\frac{T_e}{2\pi m_e}}\, n_\infty \exp\left(\frac{e\Phi(R)}{T_e}\right). \tag{5.21}$$

Thus, the Dirichlet boundary conditions are derived as

$$\Phi\bigg|_{r=+\infty} = 0, \qquad \Phi(R) = \frac{T_e}{e} \ln\left(\frac{\hat{R}^2}{R^2}\sqrt{\frac{2\pi m_e}{m_i}}\right). \tag{5.22}$$

Due to the unknown constant $\hat{R}$, the differential equation cannot be directly solved. Instead, such an ordinary differential equation boundary value problem is converted into an initial problem, which is often tackled by the shooting method in numerical analysis. As is suggested by its name, the shooting method is denoted by the analogy with the target shooting. By changing the initial value, the first boundary condition, the "shooting trajectory" is adjusted iteratively until the solution is found to satisfy the other boundary condition.

For any set of parameters, the unknown $\hat{R}$ can be found by means of this method. As an example, the plasma parameters are given as $n_e = n_\infty = 1 \times 10^{15}\,\mathrm{m^{-3}}$ and $T_e = 3\,\mathrm{eV}$. To focus on solving the numerical problem, it is convenient to introduce dimensionless notations. The potential is normalized by T_e/e,

$$\Phi \rightarrow \frac{T_e}{e}\Phi, \tag{5.23}$$

and densities in terms of the reference density n_∞,

$$n_e \rightarrow n_\infty n_e, \qquad\qquad\qquad n_i \rightarrow n_\infty n_i. \qquad (5.24)$$

and the spatial coordinate in terms of the Debye length,

$$r \rightarrow \lambda_D r. \qquad (5.25)$$

Then the dimensionless differential equation can be simplified as

$$-\frac{1}{r^2}\frac{\partial}{\partial r}r^2\frac{\partial \Phi(r)}{\partial r} = \frac{\hat{R}^2}{r^2}\sqrt{-\frac{1}{2\Phi(r)}} - \exp(\Phi(r)). \qquad (5.26)$$

The potential function can be expanded as asymptotic series in the form of

$$\Phi(r) = \sum_{n=1}^{\infty} \frac{a_n}{r^n}. \qquad (5.27)$$

The coefficients a_n are determined by inserting the asymptotic series of potential into the Poisson equation, which gives the expression of $\Phi(r)$ and $\Phi'(r)$ in terms of r and $\hat{R}$.

In practice, it is harsh to set r to infinite in the numerical calculation. Hence, assuming a r_∞ representing a large distance from the probe is reasonable, which writes $r_\infty = 10R$. Starting from $\Phi(r_\infty) = 0$, if the derivative $\Phi'(r_\infty)$ is also known from a guessing $\hat{R}$, the initial value problem with two boundary conditions at $r = r_\infty$ can be solved with some numerical methods, such as the most widely used Runge-Kutta methods. Then the shooting step is executed by integrating to the other boundary at $r = R$. By repeatedly comparing the newly calculated value of $\Phi(R)$ to the given boundary condition, and adjusting the initial value, eventually, the constant $\hat{R}$ can be found when the correct boundary condition is hit.

Runge-Kutta methods: There are many ways to solve ordinary differential equations. One of the well-known numerical methods is the Runge-Kutta (RK) method. Basically, the idea is, by repeatedly substituting into the differential equation, a sequence of approximate solutions can be obtained. This concept is developed by Runge and Kutta [104, 105]. Here, for a robust and accurate approximation, the fourth-order Runge-Kutta method, sometimes known as RK4, is used for the numerical solutions of a second-order ODE.

Let $x = r$, $y = \Phi(r)$, and $z = \Phi'(r)$. A general process of the RK method requires functions

$$\begin{cases} f(x,y,z) &= \dfrac{dy}{dx} = z, \\[2ex] g(x,y,z) &= \dfrac{dz}{dx} = \dfrac{2z}{x} + \dfrac{\hat{R}^2}{x^2\sqrt{-2y}} - e^y. \end{cases} \qquad (5.28)$$

1. At the beginning of the time step, the slopes k_0, j_0 are calculated,

$$k_0 = f(x_i, y_i, z_i) \cdot h, \tag{5.29}$$
$$j_0 = g(x_i, y_i, z_i) \cdot h.$$

2. k_1, j_1 are estimates of the slope at the midpoint after half step using k_0, j_0,

$$k_1 = f(x_i + \frac{1}{2}h, y_i + \frac{1}{2}k_0, z_i + \frac{1}{2}j_0) \cdot h, \tag{5.30}$$
$$j_1 = f(x_i + \frac{1}{2}h, y_i + \frac{1}{2}k_0, z_i + \frac{1}{2}j_0) \cdot h.$$

3. Repeating last step, k_2, j_2 are estimates of the slope with k_1, j_1,

$$k_2 = f(x_i + \frac{1}{2}h, y_i + \frac{1}{2}k_1, z_i + \frac{1}{2}j_1) \cdot h, \tag{5.31}$$
$$j_2 = f(x_i + \frac{1}{2}h, y_i + \frac{1}{2}k_1, z_i + \frac{1}{2}j_1) \cdot h.$$

4. Finally, the slopes k_3, j_3 are evaluated for the calculation,

$$k_3 = f(x_i + h, y_i + k_2, z_i + j_2) \cdot h, \tag{5.32}$$
$$j_3 = f(x_i + h, y_i + k_2, z_i + j_2) \cdot h.$$

5. Thus, y_{i+1}, z_{i+1} are determined for next step.

$$y_{i+1} = y_i + \frac{1}{6}(k_0 + 2k_1 + 2k_2 + k_3), \tag{5.33}$$
$$z_{i+1} = z_i + \frac{1}{6}(j_0 + 2j_1 + 2j_2 + j_3).$$

For n simulation steps, the step size can be calculated as $h = (r_\infty - R)/n$. The algorithm converges at boundary $r = R$ when the residual value between the shooting result to the boundary condition (5.22) reaches zero.

Consequently, the steady-state of the IMRP-plasma system with the given plasma parameters can be evaluated by means of the shooting method. According to these calculations, the spatial distribution of the particles in the initialization of the simulation is determined. As is shown in Figure 5.3, the potential of the system is demonstrated. Moreover, in Figure 5.4, the ion and electron density profiles are presented, which describe the initial state before the onset of the electric signal from the IMRP.

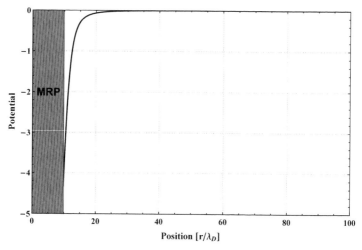

Figure 5.3: The potential of the IMRP-plasma system in steady-state is demonstrated. The simulated plasma is with a electron density $n_e = 1 \times 10^{15}\,\mathrm{m}^{-3}$ and a electron temperature $T_e = 3\,\mathrm{eV}$.

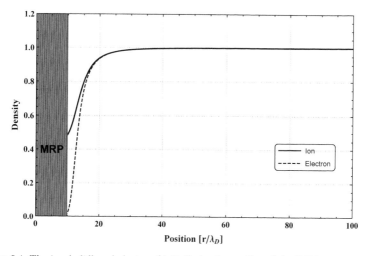

Figure 5.4: The ion (solid) and electron (dotted) density profiles of the IMRP-plasma system in steady-state are demonstrated. The simulated plasma is with an electron density $n_e = 1 \times 10^{15}\,\mathrm{m}^{-3}$ and an electron temperature $T_e = 3\,\mathrm{eV}$. According to these calculations, the initial distribution of particles is established.

As is assumed in Section 4.1, there is a total surface charge Q_S homogeneously distributed on the surface of IMRP. The system fulfills the charge equilibrium, which maintains the steady-state of the simulation. Therefore, for a plasma with N classical free point charges, the overall charge balance yields at

$$\sum_{k=1}^{\infty} q_k + Q_S = 0. \tag{5.34}$$

The surface charge is obtained from the charge density, which is from the expression

$$Q_S = \int_{\mathcal{V}} \rho(r)\,\mathrm{d}\mathcal{V} = 4\pi \int_R^{+\infty} (n_i - n_e)\, r^2 \,\mathrm{d}r. \tag{5.35}$$

Back in the study of Brinkmann [106, 107], the exact sheath thickness δ is to be obtained in the investigation of the plasma sheath, and the calculation of the surface charge is discussed. As is shown in Figure 5.5, it is possible to find a point s where the negative charges due to electrons in the interval $(R,\,s)$ correspond to the positive charges from the exceeded ion in the interval $(s,\,+\infty)$

$$4\pi \int_R^s n_e r^2 \,\mathrm{d}r = 4\pi \int_s^{+\infty} (n_i(r) - n_e(r))\, r^2 \,\mathrm{d}r. \tag{5.36}$$

After some calculations, the surface charge can be derived from the underlying expression with a given sheath thickness

$$Q_S = 4\pi \int_R^\delta n_i(r)\, r^2 \,\mathrm{d}r. \tag{5.37}$$

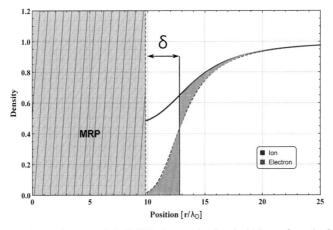

Figure 5.5: In the steady-state of the IMRP-plasma, the sheath thickness δ can be found, and the surface charge can be calculated.

Notably, in the spectral kinetic simulation, the assumption of sheath thickness is not necessary. However, it is required in the simulation of the fluid model. In further investigation, the simulation results of the kinetic and fluid model are compared, and the calculation of the input sheath thickness is according to equation (5.36). Presently in this section, focusing on the kinetic model, the exact value of surface charge can be obtained by equation (5.35) and (5.37). Besides, with the numerical ion and electron density profiles, the steady-state can be established as the initial condition of simulation. Basically, for any different set of plasma parameters, the corresponding numerical values to describe the steady-state of the system need to be evaluated. In Figure 5.6, the density profiles of the plasma for a variation of the electron temperature $T_e \in \{1, 3, 5, 7\}$ eV are demonstrated. Due to the dependence of the Debye length λ_D to the electron temperature, the spatial coordinates are normalized by the size of IMRP with the purpose that the comparison is on the same scale. The outer boundary is set as $r_\infty = 10R$, which gives the total range of the plot $r \in [R, 10R]$. In summary, depending on the investigated plasma, different initial conditions are indispensable.

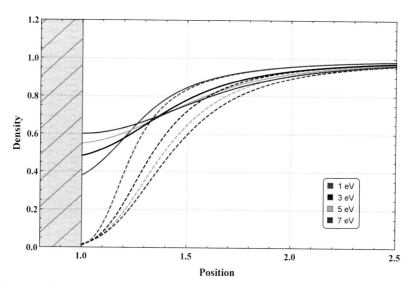

Figure 5.6: The steady states of the IMRP-plasma system with an electron density $n_e = 1 \times 10^{15}$ m^{-3} and different electron temperature $T_e \in \{1, 3, 5, 7\}$ eV are determined.

5.2.2 Boundary Conditions

In order to have a self-consistent system, the particle interactions with the surface of IMRP and the outer boundaries need to be defined. At the physical boundaries, the static surface charge is calculated using ion and electron flux expression, whereas, at the open boundaries, the Dirichlet boundary condition indicates the vanishing of potential.

During the simulation, a small number of particles can travel to the surface of the probe or out of the simulation domain. Most of them are electrons due to the relatively fast movement compared to ions. Here, two common strategies can be used at the physical boundaries, reflection or sticking of particles. For the idea of sticking of particles, the predefined surface charge is updated accordingly when negative electrons or positive ions attach to the probe. To ensure an adequate number of particles in the simulation, new ones are generated in this assumption. Contrarily, the reflection of particles indicates that the surface charge remains constant till the end of the simulation. Once the particle reaches the boundary, it is reflected with an identical speed in a random direction. In fact, only a limited number of particles are able to reach the surface. The electrons are often repelled by the strong negative surface charge, while heavy ions are barely moving. Therefore, the strategy of reflection of particles is used so that generating new particles is not obligated. Besides, regarding the outer boundary, it is of less importance to precisely monitor the inflow or outflow of the particles. As a matter of fact, the reflection of particles leads to heating phenomena due to the energy from the input signal. However, this error is minor since the input is solely an impulse in the spectral kinetic simulation.

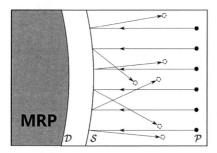

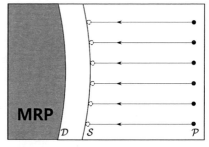

Figure 5.7: The strategy of boundary condition: particles are reflected with the same speed in a random direction after reaching the IMRP.

Figure 5.8: The strategy of boundary condition: particles are attached to the surface to update the surface charge Q_S after reaching the IMRP.

5.3 Initialization of the Spectral Kinetic Simulation

The significance of an appropriate initialization has been emphasized in this chapter. Essentially the steady-state of the system is implemented as the starting point of the simulation. Here, the collisionless case is examined for simplicity. The investigated plasma in this example is with an electron density $n_e = 1 \times 10^{15}\,\text{m}^{-3}$ and an electron temperature $T_e = 3\,\text{eV}$ (see Table 6.1). In a 3D simulation domain, each particle is assigned with a position vector $\boldsymbol{r}(x, y, z)$ and a velocity vector $\boldsymbol{v}(v_x, v_y, v_z)$ according to the particle density profiles and the electron velocity distributions. Afterward, the acceleration can be determined by force acting on each particle due to the electric fields. The input signal can be directly applied to the plasma after such initialization procedures, where the simulation time is highly reduced.

Firstly, on the basis of the afore calculated density profiles, the distribution of particles in the spatial domain can be determined. Specifically, with the given number of ions in the simulation, the number of electrons and the static surface charge can be derived from equation (5.36) and (5.37), which fulfills the charge equilibrium (5.34). It is straightforward to specify the initial position of electrons and ions, respectively. The complete simulation domain is divided into subdomains with regard to r. Since the total number of electrons or ions is set, the number of particles in each interval can be calculated. That is, the percentage of particles in each subspace is calculated from the static particle density profiles. As a result, the exact amount of particles can be generated in the corresponding simulation subdomain. As shown in Figure 5.9, the initial spatial distribution of particles can be verified by comparing it with the numerical density profiles.

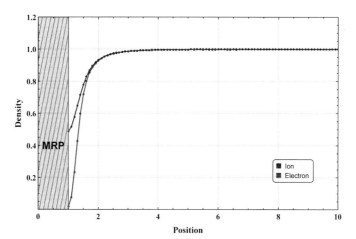

Figure 5.9: According to afore calculated results, the initial distribution of particles is established. The ion (blue) and electron (red) density profiles of the IMRP-plasma system in steady-state are demonstrated.

Furthermore, particles are generated with a variety of velocities in the simulation. The initial velocity of particles is to be decided, which is related to velocity distributions. As is introduced, the distribution function is Maxwellian (5.20), which indicates electrons are in thermodynamic equilibrium. It is often convenient to express the distribution functions in terms of energy ϵ [4]. In fact, in plasma modeling, the electron energy distribution function (EEDF), f_{EEDF}, plays an important role, the analysis of which provides a profound understanding of the plasma generation and heating mechanism [108]. The integrals of EEDF over energy can be evaluated to obtain the electron density

$$n_e = \int_0^{\infty} f_{\text{EEDF}}(\epsilon)d\epsilon. \tag{5.38}$$

The explicit Maxwell energy distribution is

$$f_\epsilon(\epsilon) = 2n_e \sqrt{\frac{\epsilon}{\pi}} \left(\frac{1}{T_e}\right)^{3/2} \exp\left(\frac{\epsilon}{T_e}\right). \tag{5.39}$$

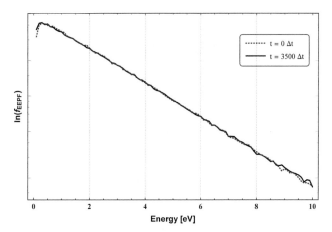

Figure 5.10: Simulated EEPFs for $\Delta u = 0$ at $t = 0\,\Delta t$ (dotted red line) and $3500\,\Delta t$ (solid black line): Maxwell distribution is verified at the initialization and remains Maxwellian.

Normally, the electron energy probability function, f_{EEPF}, is defined to better analyze the plasma process, which is obtained from EEDF divided by $\sqrt{\epsilon}$. As a unique feature of the EEPF of Maxwell distribution, the semi-logarithmic plot of EEPF is a linear function. In Figure 5.10, at $t = 0\,\Delta t$, the EEPF of the steady-state (red dotted line) verifies the initial Maxwellian distribution of electrons in the simulation. Moreover, at $t = 3500\,\Delta t$, an identical EEPF at the end of the simulation (solid black line) is captured, which indicates that the plasma remains Maxwellian. Here, the quantities are normalized in the simulation. However, it is convenient to convert the energy into unit (eV). Notably, no signal is applied in this example ($\Delta u = 0$). This result shows the stability of the spectral kinetic model.

Consequently, the initialization of the simulation is necessary to improve the efficiency and reliability of the model. In Figure 5.11, the steady-state of the system is visualized: The IMRP is surrounded by super-particles, including ions and electrons, in the complete simulation domain $r \in [R, 10R]$.

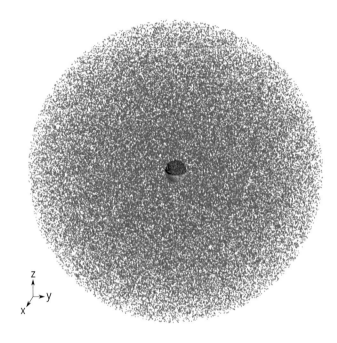

Figure 5.11: Virtualization of the simulation: the particles are spatially generated according to the predetermined density profiles. The complete simulation domain is demonstrated.

6 Implementation and Results

In preceding Chapter 5, the initialization of the simulation is appropriately established, which enables an efficient particle simulation. In the spectral kinetic simulation, an impulse signal can be directly applied to the plasma in the steady-state, then the charge difference on the electrodes is captured in the time domain. By the Fourier transform, the impulse response can be expressed in the frequency domain. Therefore, one single simulation can cover the entire frequency domain. In fact, the calculation in the frequency domain is time-consuming regarding the convergence of the sequence of periodic functions.

In this chapter, the simulation results are analyzed in detail. Firstly, the assumed energy distribution and the linearity of the system are verified to ensure the plausibility and validity of the simulation. Then the comparison of the results in the kinetic model and fluid model confirms the importance of kinetic effects in the low-pressure plasmas, which emphasizes the suitability of the proposed kinetic approach. Especially in the collisionless case of kinetic simulation, the energy losses are solely due to kinetic effects, which can be clearly observed.

Furthermore, the parameter study of the spectral kinetic model is investigated. The plasma with a variation of electron density and electron temperature is simulated. Depending on the input plasma parameters, the corresponding resonance frequency and the half-width $\Delta\omega$ are recorded for the analysis. The relation between the electron density and the resonance frequency can be evaluated. In [47, 49, 50], the resonance frequency can be predicted from different electron densities in the CST simulation via the cold plasma model. Therefore, a similar prediction is expected. Besides, a particular feature of the spectral kinetic model is to capture the kinetic effects by the half-width from the simulated resonance curves. The proposed formula for a decade in [47, 54] to describe the relation between the electron temperature and the energy loss can be clarified in this work.

Lastly, the electron-neutral collision is included in the proposed simulation, of which the results can be compared with the experiments. Specifically, in [109,110], the measurement of the MRP-plasma system is realized within a cylindrical double inductively coupled plasma reactor. The collision frequency is controlled by the adjustment of the gas pressure. In [74], the gas pressure can be directly converted from the elastic electron-neutral in certain relations depending on the investigated plasma. Hence, the simulation results can be conveniently verified. This chapter is reproduced from [18] under a Creative Commons Attribution (CC BY) license.

6.1 Simulation Results

In the following, the first simulation results are presented. For an understandable analysis, it is useful to introduce dimensionless notation. As stated in Section 5.2.1, the particle density n_∞ is assumed for the plasma where quasi neutrality holds. Regarding the basis of units, a length scale R is taken as the radius of the probe, and a time scale is defined as the inverse of the plasma frequency ω_{pe}.

$$
\begin{aligned}
\boldsymbol{r} &\rightarrow R\boldsymbol{r}, & m &\rightarrow m_{\mathrm{e}}m, & n &\rightarrow n_\infty n, & (6.1)\\
q &\rightarrow eq, & t &\rightarrow \omega_{\mathrm{pe}}^{-1}t, & \nu &\rightarrow \omega_{\mathrm{pe}}\nu.
\end{aligned}
$$

In Chapter 4, the complete spectral kinetic model is introduced. By using the Green's function method, the formal solution of the Poisson equation is established. Taking advantage of the electrical antisymmetry and the geometrical symmetry of the ideal model, the proposed simulation scheme is simplified to focus on the dipole mode. Eventually, the modified model is summarized in Section 4.3.4. With the simplified Green's function (4.86), for a random particle k, the equations of motion are

$$
\frac{\mathrm{d}\boldsymbol{r}_k}{\mathrm{d}t} = \boldsymbol{v}_k, \tag{6.2}
$$

$$
m_k\frac{\mathrm{d}\boldsymbol{v}_k}{\mathrm{d}t} = -\frac{1}{\varepsilon_0}\sum_{\substack{i=1\\i\neq k}}^{N} q_k q_i\,\nabla_k \tilde{G}(\boldsymbol{r}_k,\boldsymbol{r}_i) - q_k\nabla_k\frac{Q_S}{4\pi\varepsilon_0 r_k} - q_k\Delta u(t)\nabla_k\Delta\Psi(\boldsymbol{r}_k). \tag{6.3}
$$

In the spectral kinetic simulation, the charge difference on the electrodes (4.88) is defined as the response of the system is

$$
\Delta Q(t) = -\sum_{i=1}^{N} q_i\Delta\Psi(\boldsymbol{r}_i) + C_{\mathrm{vac}}\Delta u(t). \tag{6.4}
$$

Depending on the number of super-particles, a parameter $\hat{N}$ is defined as the expected number of super-particles in a unit cube. Therefore, the equations of motion are invariant against rescaling the number of particles to the super-particles. Taking (6.1) into account, the normalized system is derived as

$$
\frac{\mathrm{d}\boldsymbol{r}_k}{\mathrm{d}t} = \boldsymbol{v}_k, \tag{6.5}
$$

$$
m_k\frac{\mathrm{d}\boldsymbol{v}_k}{\mathrm{d}t} = -\frac{1}{\hat{N}}\sum_{\substack{i=1\\i\neq k}}^{N} q_k q_i\,\nabla_k \tilde{G}(\boldsymbol{r}_k,\boldsymbol{r}_i) - q_k\frac{1}{\hat{N}}\nabla_k\frac{Q_S}{4\pi r_k} - q_k\Delta u\nabla_k\Delta\Psi_{l=1}(\boldsymbol{r}_k). \tag{6.6}
$$

Table 6.1: Plasma parameters in the first simulation

Parameter	Value	Unit
n_e	1×10^{15}	m^{-3}
m_e	9.1094×10^{-31}	kg
m_i	6.6335×10^{-27}	kg
T_e	3	eV
T_N	300	K
ν_{col}	0	s^{-1}
N^s	999660	–

Basically, a numerical algorithm is defined to describe the interaction of plasma around the IMRP. According to Section 5.1, the output of the system linearly depends on the present and past value of the input ($t \leq t_0$) in such a causal system. In the simulation, the asymmetric part $\Delta u(t)$ is defined as the signal of the ideal MRP. The symmetric part $\bar{u}(t)$ is canceled out due to the probe is electrically antisymmetric. To mimic a behavior similar to an impulse to all the particles, the input $\Delta u = 1$ is solely applied at the first step $t = t_0$. Then the signal from the probe disappears ($\Delta u = 0$) from $t > t_0$. The normalized impulse response is recorded from

$$\Delta Q(t) = -\frac{1}{\tilde{N}} \sum_{i=1}^{N} q_i \Delta \Psi(\boldsymbol{r}_i). \tag{6.7}$$

The parameters of the investigated plasma are listed in Table 6.1, which is applied in (6.1) for the normalization. Taking argon plasma as an example, the ion mass is determined as $m_i = 6.6335 \times 10^{-27}$ kg, which is normalized by the electron mass in the simulation. The background temperature T_N is given as the room temperature $300\,K$, which is approximately $0.026\,eV$. To capture the pure kinetic effects, the elastic electron-neutral collision is not initially considered.

In the first example, the number of loaded ions is set to 5×10^5, and the electron density is assumed as $T_e = 3\,eV$. According to the numerical density profiles, the ratio of ion to electron can be derived, which decides the number of electrons. Then the total number of super-particles is determined. As stated, the total number of particles and the surface charge fulfill the charge equilibrium (5.34). Therefore, the surface charge Q_S is calculated from equation (5.35) and (5.37), which is treated as a constant in the simulation.

Regarding the size of the time step, similar to the criteria in the PIC simulation, Δt is required to be smaller than $2\omega_{pe}^{-1}$. Normally, a more restrictive condition $\Delta t = 0.01\omega_{pe}^{-1}$ is considered to avoid the nonphysical trajectories of super-particles. That is, there is no dependence on the time step to the simulation results.

To ensure the reliability of the initialization, the simulation with $\Delta u = 0$ is executed for 3500 simulation steps. The starting point of the simulation is the steady-state of the IMRP-plasma model. Then the discrete simulation results are recorded. Since the probe is switched off, the status of the model remains steady. This is compared to the case with an input signal for the same plasma. In Figure 6.1, the comparison is presented: the charge on the electrodes ΔQ for the case without any input (red) holds a status close to static. Contrarily, for the case with an impulse (black), the output in the form of an oscillation plot in the time domain can be observed. To sum up, the steady-state is correctly implemented in the simulation. Besides, with a "kick" to the plasma, the oscillation is the response of the system for the same investigated plasma.

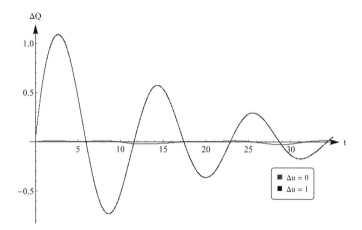

Figure 6.1: Comparison of the simulated charge difference on the electrodes ΔQ: The system retains the static condition for the case $\Delta u = 0$, whereas damping phenomena are observed in the resonance for $\Delta u = 1$

As afore introduced in Section 5.2, the electrons follow the Maxwell distribution function in the spectral kinetic simulation, which can be seen as an appropriate assumption to simulate a scenario occurring on a short time scale. In Figure 5.10, the energy distribution for case $\Delta u = 0$ is confirmed as Maxwellian from the electron energy probability function (EEPF) at the starting time step $t = 0\,\Delta t$, and the same distribution is observed at $t = 3500\,\Delta t$. The corresponding output of the system ΔQ (red) is shown in Figure 6.1. Comparably, for the situation where a signal is applied to the system ($\Delta u = 1$), the energy distribution is also investigated. In Figure 6.2, the EEPFs at $t = 0\,\Delta t$ (red dotted line) and $t = 3500\,\Delta t$ (blue solid line) are demonstrated. Similarly, the typical behavior of the Maxwell distribution function can be ascertained in the semi-logarithmic plot of EEPF. Notably, there is no obvious evolution of the energy distribution over the simulation time. That is, the initialized steady state of plasma is retained in the simulation. Moreover, the input signal causes negligible influences on the energy distribution of particles.

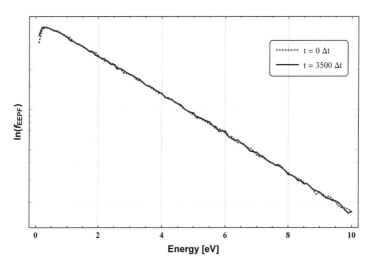

Figure 6.2: Simulated EEPFs for $\Delta u = 1$ at $t = 0\,\Delta t$ (red dotted line) and $3500\,\Delta t$ (blue solid line): Maxwell distribution is verified at the initialization and remains Maxwellian.

Moreover, to obtain the impulse response in the frequency domain, the Fourier transform is used to convert the oscillation of ΔQ from the time domain into the frequency domain. It is convenient to analyze the continuous function. Therefore, the mathematical expression can be evaluated to fit the discrete numerical results. The expansion is written in the form of

$$f(t) = \sum_{s=1}^{\infty} a_s e^{-b_s t} \sin(c_s t), \qquad (6.8)$$

where a is the amplitude, b is the damping factor, and c is the resonance frequency. The index s is a positive integer, which reads $s \in \mathbb{Z}^+$. As a matter of fact, numerous simulation results suggest that the higher modes of expansion are not necessary for finding an approximately fitted function. Thus, $s = 1$ denotes the analytical function for further investigation.

Figure 6.3 shows the example of a collisionless plasma with $T_e = 2\,\text{eV}$, $n_e = 1.0 \times 10^{15}\,\text{m}^{-3}$, and $\nu_{\text{col}} = 0$. Although plasma with different electron temperatures is simulated, a similar damped resonance behavior is captured. Then the fitted analytical function is determined as

$$f(t) = 1.135\, e^{-0.049t}\, \sin(0.509t). \qquad (6.9)$$

Notably, the damping factor $b = -0.049$ and resonance frequency $c = 0.509$ is normalized in terms of electron plasma frequency, which varies for different electron densities. The relation between the resonance frequency and electron density is to be discovered; however, presenting the plot in unit GHz is beneficial in further sections.

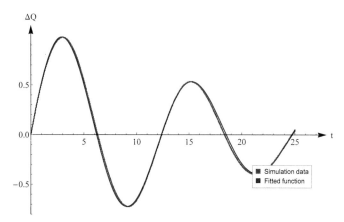

Figure 6.3: A fitted function (blue) is defined to match the simulation result (red). In this example, the investigated plasma is with the electron density $n_e = 1 \times 10^{15} \, \text{m}^{-3}$ and the electron temperature $T_e = 2 \, \text{eV}$.

Then the Fourier transform of the fitted function is determined so that the result in the frequency domain can be demonstrated. Here, in Figure 6.4, the real part of the admittance $Y(\omega)$ is derived for the characterization of the plasma: In the simulated resonance curve, the resonance frequency ω_{res} can be directly observed, and the energy loss is linked to the half-width $\Delta\omega$. The detailed parameter study is presented in the next section.

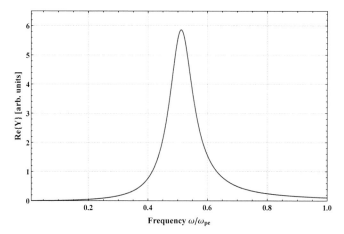

Figure 6.4: The corresponding resonance curve is obtained in the frequency domain after the Fourier transform.

Additionally, in the following example, the linearity of the plasma behavior is inspected from the simulation results. It is vital to ensure that the impulse signal solely causes linear response so that further analysis is plausible. Here, taking electron temperature at $T_e = 5\,\mathrm{eV}$ for a plasma with the same electron density $n_e = 1 \times 10^{15}\,\mathrm{m}^{-3}$, scenarios with different amplitudes of the input signal, $\Delta u = 1$ and $\Delta u = 0.5$, are simulated, respectively. The comparison of the corresponding simulated resonance curves is presented. Figure 6.5 indicates a linear dependence of the Fourier transform of the charge on the electrodes to the amplitude of the input signal. However, by deriving the admittance, it is evident that the characterization of the simulated plasma is independent of the signal from the probe. In Figure 6.6, the nearly identical resonance curves of the admittance are demonstrated, which are solely determined from the given plasma parameters.

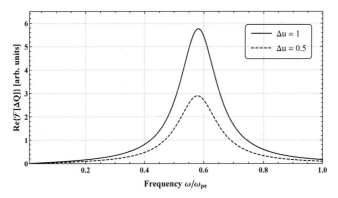

Figure 6.5: Comparison of the simulation with different input signal: $\Delta u = 1$ (solid line) leads to a higher amplitudes of the simulated resonance curve compared to $\Delta u = 0.5$ (dashed line).

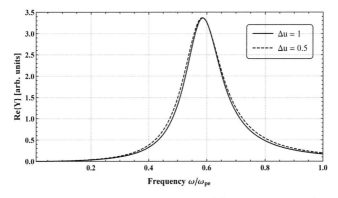

Figure 6.6: Comparison of the derived admittance with different input signal: $\Delta u = 1$ (solid line) leads to an approximately identical resonance curve compared to $\Delta u = 0.5$ (dashed line).

To summarize, the simulation for a variation of the electron temperature $T_e \in \{2,3,5\}$ eV at a constant electron density $n_e = 1 \times 10^{15}\,\mathrm{m}^{-3}$ is investigated. The elastic electron-neutral collisions are not considered in this section for simplicity. Basically, the assumptions of the spectral kinetic model are verified in the simulation results, such as the initialized steady state of the plasma. Besides, the electron energy distribution is verified at different time steps, and the linearity of the system is confirmed. Importantly, a time-varying charge on the electrodes ΔQ is recorded, where the resonance behavior is observed in the results of the spectral kinetic simulation. As expected, damping phenomena are captured in the simulated oscillations. It is interpreted as the kinetic effects: The energy losses are caused by the escape of the free particles from the influenced domain.

Furthermore, by employing a fitted function, the system output ΔQ can be conveniently converted into the frequency domain via Fourier transform, where the admittance Y of the system can be derived for the characterization of the plasma. Here, the comparison of the presented examples is presented: In Figure 6.7, the simulated resonance curves of the real part of admittance are demonstrated, where the resonance frequency ω_{res} and the half-width $\Delta\omega$ can be obtained [8]. In a nutshell, the resonance frequency is expected to be determined by the electron density n_e. However, the slight shift of ω_{res} for a variation of electron temperature appears in the simulation results. It is caused by the different sheath thicknesses of the investigated plasma. Apparently, the kinetic approach is more realistic physically since in the cold plasma model sheath thickness is solely an approximation [77]. Additionally, the increase of electron temperature results in a greater half-width of the resonance curve, the possible dependence of $\Delta\omega$ to electron temperature is to be further investigated. The corresponding parameter study is discussed in subsequent sections.

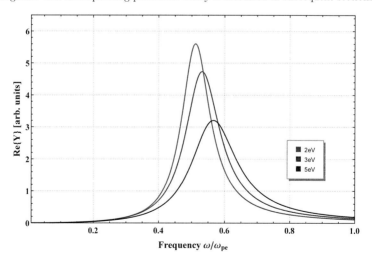

Figure 6.7: Simulation results of the collisonless spectral kinetic simulation of the IMRP for a variation of the electron temperature T_e at a constant electron density $n_e = 1 \times 10^{15}\,\mathrm{m}^{-3}$.

6.2 Resonance Frequency and Electron Density

In this section, simulations in the dipole mode with varying electron densities are ana-lyzed. The comparison between the results from the kinetic model and the fluid model is presented. In fact, the relation between the resonance frequency ω_{res} and the electron density n_{e} in the MRP system is first discussed in [17, 48]. The cold plasma model is utilized to derive an analytical expression of the admittance (2.74) and the resonance frequency (2.79), which characterizes the probe response. The details are shown in 2.2.6: The MRP-plasma system is treated as equivalent lumped element circuits, where reso-nance modes are represented by an infinite number of resonant circuits. However, the importance of the dipole mode has been confirmed. Hence, focusing solely on the first resonance peak is sufficient for the investigation. Since the collision frequency is irrele-vant to the resonance frequency, $\nu_{\mathrm{col}} = 0$ is assumed in the kinetic simulation to have the collisionless case.

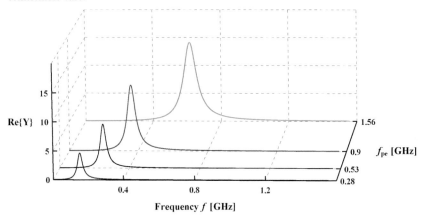

Figure 6.8: Simulated resonance behavior of the IMRP for $f_{\mathrm{pe}} \in \{0.28, 0.53, 0.9, 1.56\}$ GHz, corresponding to a variation of the plasma density between $1 \times 10^{15}\,\mathrm{m}^{-3}$ to $3.0 \times 10^{16}\,\mathrm{m}^{-3}$.

Apparently, in the cold plasma model, the electron plasma frequency can be obtained by adjusting the electron density, then the resonance frequency is evaluated. Similarly, in the kinetic model, the resonance frequency is directly determined from the simulated resonance curve. To be more specific, for a rising electron density, a rising resonance frequency is observed. This relation is numerically obtained for a variation of the given electron density. In Figure 6.8, the simulation results of the collisionless case with different electron densities are presented. Here, the electron temperature is set as a constant $T_{\mathrm{e}} = 3\,\mathrm{eV}$, and the electron densities are considered in $10^{16}\,\mathrm{m}^{-3}$: 0.1, 0.35, 1.0, 3.0. With the expression $\omega_{\mathrm{pe}} = 2\pi f_{\mathrm{pe}}$, the resulting resonance curves for the corresponding electron plasma frequencies $\omega_{\mathrm{pe}} = 2\pi \cdot \{0.28, 0.53, 0.9, 1.56\} \cdot 10^9\,\mathrm{s}^{-1}$ are recorded, including the information of the resonance frequency and the normalized admittance. The increase of the resonance frequency f_{res} is demonstrated for a rising electron density n_{e}.

Moreover, the kinetic simulation with the electron density in a broader range between $3.5 \times 10^{14}\,\mathrm{m}^{-3}$ and $1.0 \times 10^{17}\,\mathrm{m}^{-3}$ are investigated, which corresponds to a variation of the electron plasma frequency f_{pe} from 170 MHz to 2.84 GHz. In Figure 6.9, a collection of resonance curves of dipole mode at $T_{\mathrm{e}} = 3\,\mathrm{eV}$ is shown. A higher amplitude refers to a stronger electric current between the electrodes, and importantly, the simulation results exhibit a clear tendency to increase in resonance frequencies f_{res} due to the rising electron density.

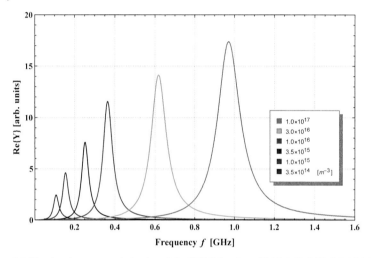

Figure 6.9: Simulated resonance behavior of the IMRP for $f_{\mathrm{pe}} \in \{0.17, 0.28, 0.53, 0.9, 1.56, 2.84\}$ GHz, corresponding to a variation of the plasma density between $3.5 \times 10^{14}\,\mathrm{m}^{-3}$ to $1 \times 10^{17}\,\mathrm{m}^{-3}$.

The reliability of the simulation results of the spectral kinetic model can be verified by the comparison of the fluid model. In the cold plasma model, the real part of the admittance Y is derived from the analytical model, and the exact value of resonance frequency can be determined with an appropriate assumption of the sheath thickness. Notably, in [49, 50], a similar prediction of the resonance frequency from the electron density via the cold plasma model is achieved within 3D-electromagnetic field simulations using CST Microwave Studio. It ensures the validity of the results based on the cold plasma model.

To be more specific, the complex admittance of the system is calculated from

$$Y(\omega) = i\omega C_{\mathrm{vac}} + \sum_{l=1}^{\infty} C_l \left(\frac{1}{i\omega} + \frac{i\omega + \nu}{\eta_l^2 \omega_{\mathrm{pe}}^2} \right)^{-1}, \tag{6.10}$$

and the resonance frequency of the dipole mode is

$$\omega_{\mathrm{res},1} = \sqrt{\frac{2}{3}\left(1 - 0.82\left(1 + \frac{\delta}{R}\right)^{-3}\right)}\, \omega_{\mathrm{pe}}. \tag{6.11}$$

The electron-neutral collision frequency ν_{col} is required to derive the admittance (6.10) in the cold plasma model. Therefore, it is complicated to compare the admittance for collisionless cases. In [47], a compensated frequency is defined to mimic the kinetic effects. However, it is possible to only compare the resonance frequency for different electron densities between the kinetic model and the cold plasma model. Here, the sheath thickness is assumed as $\delta = 3\lambda_D$ in (6.11). As a result, the calculated resonance frequency in the cold plasma model (black circles) and the simulated resonance frequency in the kinetic model (black dots) for the same electron plasma frequency are presented in Figure 6.10. A reasonable agreement can be achieved. Although the results for higher electron plasma frequencies show a divergence in the resonance frequency, the resonance behavior is comparable. In fact, considering that the assumption of a sheath thickness is not necessary for this self-consistent particle-based model, the spectral kinetic scheme can be seen as a reliable approach to predicting the resonance frequency based on the electron density.

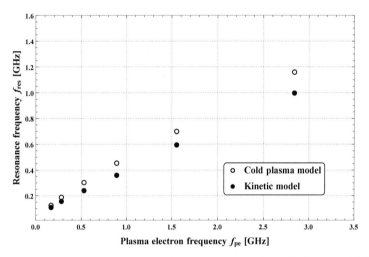

Figure 6.10: Comparison of the resonance frequencies between the cold plasma model (black circles) and the kinetic model (black dots) for a variation of the electron plasma frequencies between 0.17 and 2.84 GHz.

Previously, in Figure 6.7, a slight shift of the resonance frequency has been observed for plasmas with the same electron density and different electron temperatures. This poses the problem of whether the weak dependence of resonance frequency to electron temperature causes a failure in the prediction of electron density. As a matter of fact, the often considered electron temperature in technical plasmas is in a range between 1 eV and 7 eV [74], which indicates the limits of the electron temperature. In Figure 6.11, the simulation results for electron temperature $T_e \in \{2, 3, 5\}$ eV and electron plasma frequency $f_{pe} \in \{0.28, 0.53, 0.9\}$ GHz are presented. As expected, a positive correlation between the simulated resonance frequency and the given electron plasma frequency can be determined. Besides, the increase in electron temperature leads to a moderately higher resonance frequency. However, in Figure 6.12, compared to the results in the range of the investigated electron frequency, the shift caused by the electron temperature to the resonance frequency is negligible for the investigated plasmas.

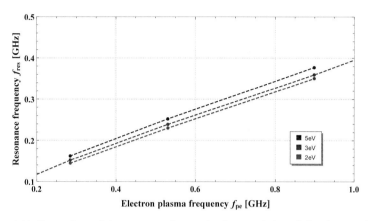

Figure 6.11: Comparison of the resonance frequencies for a variation of the electron plasma frequencies $f_{pe} \in \{0.28, 0.53, 0.9\}$ GHz and electron temperatures $T_e \in \{2, 3, 5\}$ eV.

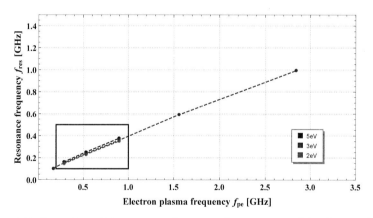

Figure 6.12: Comparison of the resonance frequencies for a variation of the electron plasma frequencies f_{pe} between 0.17 and 2.84 GHz, the influence of electron temperature is barely observable.

6.3 Half-width and Electron Temperature

As aforestated, it is possible to define a formula for the determination of the electron temperature from the half-width of the resonance curve, which can also be used to derive the electron density from the resonance frequency. Hence, only one measure is needed to obtain the plasma parameters for the characterization of plasma. The half-width $\Delta\omega$, also known as the full width at half maximum (FWHM), describes the damping of the IMRP-plasma system. It can be determined to evaluate the effective collision frequency ν_{eff}, which writes

$$\nu_{\text{eff}} = \nu_{\text{col}} + \nu_{\text{kin}}, \tag{6.12}$$

where the rate of momentum transfer from electron to neutrals is described by ν_{col}, and ν_{kin} imitates the influence of the kinetic effects. In general, the elastic collisions between electrons and neutrals are dominant due to the low ionization degree in technical plasmas. However, in the low-pressure regime, the kinetic effects dominate the total energy loss. The particles are deflected by the field of the IMRP, and the energy is transported out of the perturbed domain of the plasma. In order to reveal the influence of the kinetic effects on $\Delta\omega$, the spectral kinetic simulation of the collisionless case is investigated. Here, the assumption of collisionless ($\nu_{\text{col}} = 0$) indicates that ν_{eff} is identical to ν_{kin}, where ν_{eff} is proportional to $\Delta\omega$. On the contrary, in the cold plasma model, the only energy loss within the plasma is due to the electron-neutral collisions ($\nu_{\text{kin}} = 0$). Therefore, if ν_{col} is given in the cold plasma model, by comparing the resonance curves with the kinetic model, the exact value of $\Delta\omega$ in the kinetic model can be determined. In Figure 6.13, the

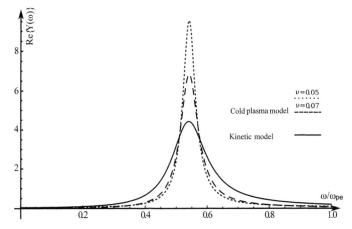

Figure 6.13: Comparison of the resonance curves between the cold plasma model and the kinetic model: the admittance can be calculated directly in the cold plasma model with collision frequency at $0.05\omega_{\text{pe}}$ (dotted line) and $0.07\omega_{\text{pe}}$ (dashed line) whereas it is derived from the charge on the electrodes in the kinetic model (solid line).

comparison of the admittance is presented [8]. The admittance of the IMRP in the kinetic model can be derived from the simulated charge difference, and its spectrum is plotted. The investigated plasma is with a electron density $1 \times 10^{15}\,\mathrm{m^{-3}}$ and electron temperature $3\,\mathrm{eV}$. To have a clear view of the comparison, the sheath thickness is chosen accordingly in the cold plasma model to match the resonance frequency of the result in the kinetic simulation. Eventually, an increasing collision frequency causes a broadening in $\Delta\omega$ and a decreasing amplitude in the cold plasma model. Notably, even in the collisionless case, the resonance curve of the kinetic model is much broader, which indicates that the fluid model is with limited validity due to the absence of the kinetic effects. Contrarily, these effects are well demonstrated in the spectral kinetic simulation.

Furthermore, the admittance of IMRP in the cold plasma model Y_{Fluid} is introduced in equation (6.10). It is expressed as the function of the sheath thickness δ, the collision rate ν, and the frequency ω itself, whereas the admittance in the kinetic model Y_{Kin} is only the function of the frequency. The curve fitting can be implemented by the method of least squares, which writes

$$I = \int_0^{\omega_{\mathrm{pe}}} \left(Y_{\mathrm{Fluid}}(\delta, \nu, \omega) - Y_{\mathrm{Kin}}(\omega) \right)^2 \mathrm{d}\omega. \tag{6.13}$$

Therefore, finding the minimal value of I provides the ν_{eff} from the evaluated ν in the cold plasma model. As shown in Figure 6.14, the curves match at $\nu = 0.128\omega_{\mathrm{pe}}$, which determines the $\Delta\omega$ in the spectral kinetic simulation.

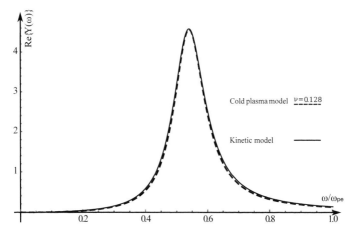

Figure 6.14: Comparison of the resonance curves between the cold plasma model (dashed line) and the kinetic model (solid line): the effective collision rate ν_{eff} can be determined by matching $\Delta\omega$ of the resonance peaks.

As stated in [47,54], a formula was speculated for over a decade, which describes a linear relation between the electron temperature to the kinetic collision frequency. With the thermal velocity of the electrons $v_{th,e}$, and an unknown length scale factor L, we can write

$$\nu_{kin} = \frac{v_{th,e}(T_e)}{L}. \tag{6.14}$$

Here, the thermal velocity can be expanded in terms of Debye length and resonance frequency as $v_{th,e} = \lambda_D(T_e)\,\omega_{pe}$. It is known that the Debye length depends on the electron temperature. Assuming a length scale $L = R$, the proposed formula can be expressed with an unknown coefficient k as

$$\Delta\omega = k \cdot \frac{\lambda_D(T_e)\omega_{pe}}{R}. \tag{6.15}$$

It indicates that $\Delta\omega$ of the simulated resonance curve is of particular interest evaluating T_e. The underlying relation is determined by altering the electron temperature in the simulation.

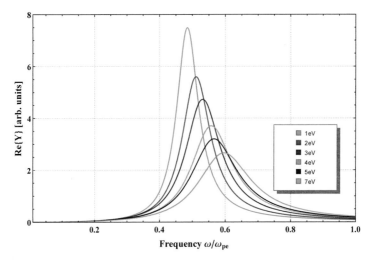

Figure 6.15: Simulation results of the collisonless spectral kinetic simulation of the IMRP for a variation of the electron temperature $T_e \in \{1, 2, 3, 4, 5, 7\}$ eV at a constant electron density $n_e = 1 \times 10^{15}\,\mathrm{m^{-3}}$.

Taking the plasma with a same electron density $n_e = 1 \times 10^{15}\,\mathrm{m^{-3}}$ as an example. In Figure 6.15, the simulated resonance curves for a variation of the electron temperature $T_e \in \{1, 2, 3, 4, 5, 7\}$ eV are presented. It shows that a rising electron temperature leads to the broadening of the resonance curve due to pure collisionless kinetic effects. Then the half-width is determined as $\Delta\omega \in \{0.075, 0.105, 0.129, 0.148, 0.167, 0.194\}\,\omega_{pe}$. Thus, this coefficient can calculated as $k = 1.264$ at this specific investigated electron density.

As shown in Figure 6.16, the calculated $\Delta\omega/\omega_{pe}$ (blue dots) can be approximated as a linear function of λ_D/R, where the coefficient k is the slope. Similarly, such a formula can be derived from the simulation results for a variation of the electron density. For the plasma with $n_e = 3.5 \times 10^{14}\,\mathrm{m^{-3}}$ and $n_e = 3.5 \times 10^{15}\,\mathrm{m^{-3}}$, simulations of $T_e = 2\,\mathrm{eV}$, $3\,\mathrm{eV}$, and $5\,\mathrm{eV}$ are executed, respectively. The results (black and red dots) agree well with the proposed formula. The coefficient yields at $k = 1.232$. Eventually, within a certain magnitude the proposed formula can be obtained as

$$\Delta\omega(T_e) = \frac{k}{R}\sqrt{\frac{T_e}{m_e}}. \tag{6.16}$$

As expected, a linear relation between the half-width $\Delta\omega$ and the electron temperature T_e is verified in the parameter study of the spectral kinetic model. Strictly speaking, there is a slight deviation due to the weak dependence of the electron density on the slope k. Therefore, a modified formula is expected to cover such second-order effects in a broader range of electron density.

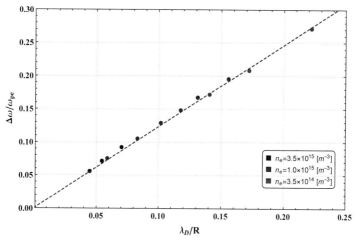

Figure 6.16: Evaluated FWHM $\Delta\omega$ of the simulated resonance curves with different plasma parameters.

However, with the prediction of $\Delta\omega$ from the investigated plasma using spectral kinetic simulation, in practice a lookup table can be created to determine the desired electron temperature from the measured half-width alternatively. Specifically, the simulation results for the electron density n_e in the range from $3.5 \times 10^{14}\,\mathrm{m^{-3}}$ to $3 \times 10^{16}\,\mathrm{m^{-3}}$ are

Table 6.2: An example of a lookup table of $\Delta\omega$ in $10^8\,s^{-1}$

n_e \ T_e	2 eV	3 eV	5 eV
$3.5 \times 10^{14}\,\mathrm{m}^{-3}$	1.815	2.167	2.831
$1.0 \times 10^{15}\,\mathrm{m}^{-3}$	1.873	2.302	2.975
$3.5 \times 10^{15}\,\mathrm{m}^{-3}$	2.256	2.750	3.558
$1.0 \times 10^{16}\,\mathrm{m}^{-3}$	3.013	3.588	4.479
$3.0 \times 10^{16}\,\mathrm{m}^{-3}$	3.494	4.231	5.212

demonstrated in Table 6.2, where the half-width $\Delta\omega$ of different electron temperature is evaluated. As observed, a linear relation between $\Delta\omega$ to T_e for each electron density can be assumed. According to the measured resonance frequency, the electron density is obtained. Then the desired corresponding electron temperature can be calculated from the measured half-width due to the proportionality.

6.4 Validation of the Simulation Results

To validate and verify the spectral kinetic model, a more physically realistic configuration in simulation is imperative. In actuality, it is difficult to measure the plasma parameters in collisionless cases. Therefore, for a complete investigation of the IMRP-plasma system, besides kinetic effects, the elastic collisions between electrons and neutral atoms need to be considered in the total energy loss. The relation between the collision frequency and resonance width is discussed in [111], where the plasma is considered via the fluid model and the absence of kinetic effects is covered by a compensated electron frequency.

As introduced in Section 3.4, the evaluation of the electron-neutral collision processes can be accomplished probabilistically in the spectral kinetic simulation. The Monte Carlo collision model is coupled with the proposed kinetic model for describing the collisional interaction. Conventionally, the Monte Carlo model is often utilized in statistical physics, where the unknown physical values can be estimated by using the principles of inferential statistics. In this work, instead of considering the collision with classical mechanics in particle simulation, the description of the collision is converted into the solution of a probabilistic problem. Particularly, all trajectories of simulated particles are integrated simultaneously at the same time step, which ensures the compatibility of using the Monte Carlo collision method. Depending on the given collision frequency, the probability of the occurrence of electron-neutral collision is determined for each particle in a time step. Repeatedly, the implementation of collisions is realized in each simulation cycle.

With the consideration of the elastic collisions, the spectral kinetic simulation is executed. Figure 6.17 shows the comparison of the simulated resonance curve for varying electron-neutral collision frequencies $\nu \in \{0, 0.005, 0.01, 0.015, 0.03\}\,\omega_{\mathrm{pe}}$. In this case, the plasma is with electron density $n_{\mathrm{e}} = 3 \times 10^{16}\,\mathrm{m}^{-3}$ and electron temperature $T_{\mathrm{e}} = 3\,\mathrm{eV}$. The increase of electron-neutral collision frequency results in a broadening of the simulated resonance curve. It is introduced that $\Delta\omega$ represents the damping in the system, which consists of the influence of the kinetic effects (ν_{kin}) and the energy loss caused by the elastic collisions (ν_{col}). As expected, the kinetic effects play an essential role in the low-pressure regime. The corresponding half-width is numerically determined as $\Delta\omega \in \{0.424, 0.492, 0.549, 0.61, 0.81\}\,\omega_{\mathrm{pe}}$. Therefore, the total energy loss can be feasibly predicted using the spectral kinetic model.

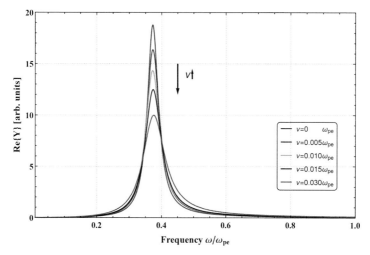

Figure 6.17: Simulated resonance curve with varying collision frequencies for the plasma with $T_{\mathrm{e}} = 3\,\mathrm{eV}$ and $n_{\mathrm{e}} = 3 \times 10^{16}\,\mathrm{m}^{-3}$.

Notably, it is observable that the resonance frequency is closely retained in spite of the increase of the electron-neutral collision frequency. That is, there is no dependence of the collision frequency on the electron density. In any measured resonance curve, the electron density can be determined without any information on collisions.

Eventually, the validation of the simulation results is completed via a comparison with experiments, where the spatially resolved *in-situ* measurements are performed. Specifically, the measurements of the MRP-plasma system with the same investigated plasma parameters are given in [75, 76]. The electron density and the electron temperature are fixed as constants: $n_{\mathrm{e}} = 3 \times 10^{16}\,\mathrm{m}^{-3}$ and $T_{\mathrm{e}} = 3\,\mathrm{eV}$. The half-width $\Delta\omega$ is evaluated for a variety of pressure from $3\,\mathrm{Pa}$ to $20\,\mathrm{Pa}$, which is previously demonstrated in Figure 1.6.

The setup of the experiment is a cylindrical double inductively coupled plasma (DICP) reactor with a diameter 40 cm and height 20 cm. Approximately, the volume of the reactor is 25 l. Within the volume, the movable probe can be placed in the center of discharge horizontally and vertically. Then the plasma parameters of argon gas are validated by reference measurements with a Langmuir probe in the APS3-system [112], where the requirements to evaluate the probe system are detailed in [113, 114].

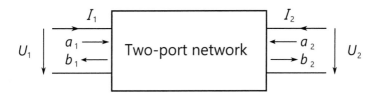

Figure 6.18: schematic diagram of a 2-port network: The reflection coefficient of the scattering parameter can be obtained.

The plasma process of MRP is controlled via excitation power and gas pressure. The resonance frequency is held constant by fixing the excitation power, and the comparable collision frequency is achieved by the adjustment of the gas pressure [109, 110]. A network analyzer is employed to measure the complex reflection coefficient. Here, the MRP belongs to the family of one-port concepts. It indicates that the influence of the plasma on the radiated electromagnetic radiation is investigated using only one port on the network analyzer. In Figure 6.18, the schematic diagram of a 2-port network is presented. The S-parameter matrix for the 2-port network is commonly used for the analysis of the reflection coefficient in a system

$$\begin{pmatrix} b_1 \\ b_2 \end{pmatrix} = \begin{pmatrix} S_{11} & S_{12} \\ S_{21} & S_{22} \end{pmatrix} \begin{pmatrix} a_1 \\ a_2 \end{pmatrix}. \tag{6.17}$$

The reflection coefficient of the scattering parameter, also known as S_{11}, is determined for a precise interpretation of the measurement, which describes the ratio of the incident wave a_1 to the reflected wave b_1 of MRP,

$$S_{11} = \left.\frac{b_1}{a_1}\right|_{a_2=0}. \tag{6.18}$$

Afterward, the real part of the admittance Y is transformed from the obtained input reflection S_{11}.

Since the pressure is decided by collision frequency in the spectral kinetic simulation, to compare the simulation results and the measurements, converting the collision frequency to the pressure is required. As discussed earlier, collisions are the fundamental process in the plasma. The general discussion is given by [4]

$$\nu_{\text{col}} = \frac{p_{\text{gas}}}{k_B T_N} \cdot K(T_e), \tag{6.19}$$

where T_N is the temperature of the neutral background, p_{gas} represents the gas pressure, and k_B is the Boltzmann's constant. $K(T_e)$ states that the elastic collision frequency ν_{col} is a function of T_e. According to [74], the elastic collision frequency in argon for T_e between 1eV and 7eV is approximated by

$$\nu_{\text{col}} = \frac{p_{\text{gas}}}{k_B T_N} \exp\left[-31.388 + 1.609\ln\left(\frac{T_e}{\text{eV}}\right) + 0.062\ln\left(\frac{T_e}{\text{eV}}\right)^2 - 0.117\ln\left(\frac{T_e}{\text{eV}}\right)^3\right]. \tag{6.20}$$

Therefore, the electron-neutral collision frequencies $\nu \in \{0, 0.005, 0.01, 0.015, 0.03\}\,\omega_{\text{pe}}$ can be inserted into equation (6.20) to calculate the corresponding gas pressure, which is obtained as $p_{\text{gas}} \in \{0, 1.607, 3.214, 4.821, 9.64\}\,\text{Pa}$. Consequently, in Figure 6.19, the comparison of the half-width from the experiment (blue dots) and the simulation results of the spectral kinetic model (red dots) is demonstrated, in addition to the analytical results of the fluid model (black dashed line).

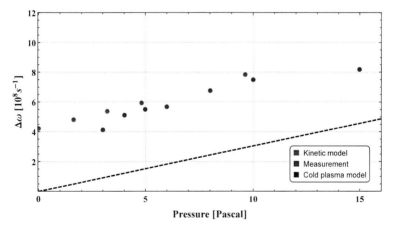

Figure 6.19: Comparison of the half-width $\Delta\omega$ between the measured spectra, the simulated results in the spectral kinetic model, and the electron-neutral collision frequency in the cold plasma model.

The results for different pressure are summarized: It shows that $\Delta\omega$ of the measured spectra is comparable to the results of the simulated resonance curve in the kinetic model.

Besides, the results based on the cold plasma model can be derived directly from the given collision frequency due to the absence of the kinetic collision frequency. Notably, it is challenging to measure $\Delta\omega$ in low-pressure plasmas. The kinetic results indicate the possible errors in the measurements, especially for the collisionless case. Although the results in higher pressure are accessible in spectral kinetic simulation, the prediction in the range from 0 to 10 Pa is sufficient to cover the limited performance of fluid approaches. In particular, the pure kinetic effects are captured in the collisionless case, which is of significant importance for the understanding of plasma behavior in a low-pressure regime. Furthermore, a good agreement between the kinetic simulation and the measurement confirms the suitability and reliability of the proposed spectral kinetic model.

7 Conclusion und Outlook

7.1 Conclusion

The multipole resonance probe is designed based on active plasma resonance spectroscopy, which is one of the industry-compatible approaches to plasma diagnostics. This work is concerned with a kinetic simulation to describe the energy transport in an MRP-plasma system. Although many studies of MRP have been proposed for over a decade, the offset caused by the limitation of the cold plasma model has not been completely explained. This research aimed to identify the influence of the missing kinetic effects on the resonance structure of technological plasmas in the low-pressure regime. From the theoretical fundamentals of plasma modeling in Chapter 2, the difference between the kinetic and fluid models of plasmas was clarified. Despite the high efficiency, the cold plasma model can not describe energy transport microscopically, which indicates the necessity of the kinetic description. Therefore, numerical methods, such as the particle-in-cell method, were introduced for solving the complicated particle model in Chapter 3. Due to the symmetry of the idealized MRP, the spectral kinetic scheme was proposed to reveal the kinetic behavior of plasmas, however, without employing any numerical grids. Moreover, the necessary collision was implemented in the Monte Carlo collision model for a complete investigation.

The details of the spectral kinetic model were discussed in Chapter 4: The dynamics of particles were described in the hamiltonian formalism, and a Poisson problem is solved with Green's function method. Specifically, the Green's function was given by an infinite expansion and had to be truncated to determine a spectrum of ideal MRP. Therefore, only the dipole mode was taken into consideration. The higher modes are absent due to their minor influence on the results. With an appropriate truncation of the sum of Green's function, a modified kinetic model was established for the investigation of dipole resonances.

Briefly speaking, the proposed probe-plasma model is a causal system, the linear response of the impulse signal in this system is to be recorded and analyzed. With the intention to optimize the simulation, the initial conditions were specified in Chapter 5: Due to the slow reaction of ions, the sheath formation was extremely time-consuming. Therefore, the steady-state was numerically solved in Boltzmann-Poisson equations, where the same particle density profiles and the surface charge were determined. Besides, considering that the simulated scenario occurred on a short time scale and the input signal was small, the Maxwell distribution function was integrated, and the diffuse reflection boundary condition was applied for simplicity. Eventually, the steady-state was calculated as the initialization of the simulation. The impulse signal can be directly provided from the probe at the first time step of the simulation, which drastically reduces the simulation time.

Chapter 6 finally regarded the simulation results. It is known that the calculations in the frequency domain are cumbersome regarding the convergence of a sequence of periodic functions. Hence, the charge on the electrodes was recorded in the time domain, and then this impulse response was converted into the frequency domain via a Fourier transform. That is, the output of the simulation was analyzed in both the time and frequency domains. To ensure the reliability of the simulation, the verification of the initial conditions was completed in Section 6.1. Then a comparison between the spectral kinetic and the cold plasma models was presented. Notably, damping phenomena in the oscillation and the broadening of the resonance curve in the kinetic model were obtained as expected. Additionally, the visualization of the simulation demonstrated the exact physical scenario related to the damped resonance behavior. Such effects were interpreted as the kinetic effects, which cover the energy loss due to the escape of the free particles from the influenced domain within the plasma. The influence of those effects was particularly emphasized in the low-pressure regime.

Furthermore, the possibility of simultaneously evaluating the electron density n_e and the electron temperature T_e from the resonance curve was explored in a parameter study of the spectral kinetic model in Sections 6.2 and 6.3. Since the resonance frequency ω_{res} is proportional to the plasma frequency ω_{pe}, the desired electron density can be derived from a measured ω_{res}. This prediction was confirmed by the comparable results of the cold plasma model. Besides, with the same measured resonance curve, the electron temperature can also be calculated from the half-width $\Delta\omega$. To be more specific, simulation of the collisionless cases relates to the pure kinetic collision frequency, which is of particular importance in evaluating the electron temperature. In other words, the kinetic collision frequency was expressed as a function of T_e. That is, importantly, a formula for the determination of electron temperature is proposed, which was speculated for over a decade, however, not been clarified. From the relation between half-width $\Delta\omega$ to the Debye length λ_D for a variation of electron density, the formula was verified in the spectral kinetic simulation. Lastly, in Section 6.4, the elastic collision between electrons and neutral atoms was integrated into the kinetic simulation for a complete investigation of the IMRP-plasma system. Utilizing the Monte Carlo collision model, the collisional interaction within plasmas was achieved probabilistically. Then a more realistic simulation was implemented to compare with the measurement. Eventually, the spectral kinetic simulation shows a good agreement with the experiment using MRP, which was realized in a cylindrical double inductively coupled plasma reactor. Here, the plasma parameters were validated by reference measurements with a Langmuir probe. It was shown that the captured energy losses were comparable in the pressure $p \leq 10\,\text{Pa}$, which indicates the reliability and suitability of the spectral kinetic model in low-pressure plasmas. Especially it is difficult to execute the measurement for a plasma in extremely low pressure. Consequently, the pronounced kinetic effects were intensively investigated, where the explanation for the offset in the cold plasma model was found. The questions concerning the kinetic effects were well-answered in this work.

7.2 Outlook

Although the spectral kinetic simulation showed an outstanding performance in covering the limitation of the fluid model, many aspects of more complicated physical scenarios go beyond the scope of this work. Besides, for the enhancement of accuracy in kinetic simulation, the further optimization of the proposed model is still pending.

Firstly, since particle-in-cell / Monte Carlo collision simulation is often seen as a powerful tool for describing the kinetic behavior of plasmas, the comparison between the results of PIC and spectral kinetic simulation is beneficial in evaluating the accuracy of the proposed model. Secondly, some assumptions can be considered to achieve a more realistic simulation. For instance, instead of the Maxwell energy distribution function, other energy distribution functions can be investigated accordingly, where energetic electrons play specific roles in different plasma systems, such as capacitively coupled plasmas or inductively coupled plasmas. In addition, one of the unique features of MRP is the prominent dipole mode, which is the focus of this work. However, the consideration of higher modes can be of particular interest for the understanding of the resonance structure. Besides, despite the fact that the idealized MRP model is proven to be suitable for the theoretical investigation, the influence of the holder can still be contained in further study. Then the comparison with the previous results from CST simulation of realistic MRP in [22, 51] can be drawn. Moreover, a promising subject is the parameter study of the model configuration, such as the radius of the probe and the size of the simulation domain. Then the second-order effects in the proposed formula are required to be resolved so that the weakly dependence on the electron density in the prediction of electron temperature can be diminished. Eventually, a modified formula to determine the electron temperature over a wide electron density range is expected for future developments. Furthermore, from an application perspective, the necessary effort to adapt the spectral kinetic model to some other plasma diagnostic device with a similar design is essential, for example, the plasma absorption probe. Lastly, similar to the simulation of argon plasma in this work, many other widely-used gases are to be simulated for an adequate database.

Eventually, the ultimate goal is, by integrating the accurate prediction from the spectral kinetic simulation, real-time supervision and control of the plasma process can be accomplished via the MRP system.

Bibliography

[1] I. H. Hutchinson. *Principles of Plasma Diagnostics. 2nd edition.* Cambridge University Press, Cambridge, 1975.

[2] K. Muraoka and M. Maeda. *Laser-Aided Diagnostics of Plasmas and Gases.* Institute of Physics Publishing, Bristol, 2001.

[3] I. Langmuir. The interaction of electron and positive ion space charges in cathode sheaths. *Physical Review*, 33(6):954–989, 1929.

[4] M. A. Lieberman and A. J. Lichtenberg. *Principles of Plasma Discharges and Materials Processing.* John Wiley & Sons, New York, 2005.

[5] F. F. Chen. *Introduction to Plasma Physics.* Springer Science & Business Media, Berlin, 2012.

[6] T. Styrnoll, J. Harhausen, M. Lapke, R. Storch, R. P. Brinkmann, R. Foest, A. Ohl, and P. Awakowicz. Process diagnostics and monitoring using the multipole resonance probe in an inhomogeneous plasma for ion-assisted deposition of optical coatings. *Plasma Sources Science and Technology*, 22(4):045008, 2013.

[7] L. Tonks and I. Langmuir. Oscillations in ionized gases. *Physical Review*, 33(2):195–210, 1929.

[8] J. Gong, M. Friedrichs, J. Oberrath, and R. P. Brinkmann. Kinetic simulation of the ideal multipole resonance probe. *Journal of applied physics*, 132(6):064502, 2022.

[9] L. Tonks. The high frequency behavior of a plasma. *Physical Review*, 37(11):1458, 1931.

[10] L. Tonks. Plasma-electron resonance, plasma resonance and plasma shape. *Physical Review*, 38(6):1219, 1931.

[11] H. Mott-Smith and I. Langmuir. The theory of collectors in gaseous discharges. *Physical Review*, 28(4):727, 1926.

[12] R. L. Merlino. Understanding langmuir probe curent-voltage characteristics. *American Journal of Physics*, 75(12):1078–1085, 2007.

[13] J. D. Swift and M. J. R. Schwar. *Electrical Probes for Plasma Diagnostics.* Iliffe Books, London, 1970.

[14] P. M. Chung, L. Talbot, and K. J. Touryan. *Electric Probes in Stationary and Flowing Plasmas: Theory and Application.* Springer, Berlin, 1975.

[15] N. Hershkowitz. *How Langmuir Probes Work Vol. 1.* Academic Press, Boston, 1989.

[16] F. F. Chen, J. D. Evans, and W. Zawalski. Electric probes. In *In Plasma Diagnostic Techniques, edited by Huddlestone, RH and Leornard, SL.* Citeseer, 1965.

[17] M. Lapke, T. Mussenbrock, and R. P. Brinkmann. The multipole resonance probe: A concept for simultaneous determination of plasma density, electron temperature, and collision rate in low-pressure plasmas. *Applied Physics Letters*, 93(5):051502, 2008.

[18] J. Gong, M. Friedrichs, J. Oberrath, and R. P. Brinkmann. The multipole resonance probe: Simultaneous determination of electron density and electron temperature using spectral kinetic simulation. *Plasma Sources Science and Technology*, 31(11):115009, 2022.

[19] J. Oberrath, M. Friedrichs, J. Gong, M. Oberberg, D. Pohle, C. Schulz, C. Wang, P. Awakowicz, R. P. Brinkmann, M. Lapke, T. Mussenbrock, T. Musch, and I. Rolfes. On the multipole resonance probe: Current status of research and development. *IEEE Transactions on Plasma Science*, 49(11):3293–3298, 2021.

[20] House of Plasma. Multipole resonance probe - designs. `https://house-of-plasma.com/start/applications/`. Accessed: 2021-11-11.

[21] C. Schulz and I. Rolfes. A new approach on advanced compact plasma sensors for industrial plasma applications. In *2014 IEEE Sensors Applications Symposium (SAS)*, pages 263–266. IEEE, 2014.

[22] C. Schulz, T. Styrnoll, P. Awakowicz, and I. Rolfes. The planar multipole resonance probe: Challenges and prospects of a planar plasma sensor. *IEEE Transactions on Instrumentation and Measurement*, 64(4):857–864, 2014.

[23] M. Friedrichs and J. Oberrath. The planar multipole resonance probe: a functional analytic approach. *EPJ Techniques and Instrumentation*, 5(1):1–15, 2018.

[24] R. L. Stenzel. Microwave resonator probe for localized density measurements in weakly magnetized plasmas. *Review of Scientific Instruments*, 47(5):603–607, 1976.

[25] R. B. Piejak, V. A. Godyak, R. Garner, B.M. Alexandrovich, and N. Sternberg. The hairpin resonator: A plasma density measuring technique revisited. *Journal of Applied Physics*, 95(7):3785, 2004.

[26] J. Xu, K. Nakamura, Q. Zhang, and H. Sugai. Simulation of resistive microwave resonator probe for high-pressure plasma diagnostics. *Plasma Sources Science and Technology*, 18(4):045009, 2009.

[27] A. Pandey, W. Sakakibara, H. Matsuoka, K. Nakamura, and H. Sugai. Curling probe measurement of electron density in pulse-modulated plasma. *Applied Physics Letters*, 104(2):024111, 2014.

[28] A. Arshadi and R. P. Brinkmann. Analytical investigation into the resonance frequencies of a curling probe. *Plasma Sources Science and Technology*, 25(4):045014, 2016.

[29] A. Arshadi, R. P. Brinkmann, M. Hotta, and K. Nakamura. A simple and straightforward expression for curling probe electron density diagnosis in reactive plasmas. *Plasma Sources Science and Technology*, 26(4):045013, 2017.

[30] H. Kokura, K. Nakamura, I. P. Ghanashev, and K. Nakamura. Plasma absorption probe for measuring electron density in an environment soiled with processing plasmas. *Japanese Journal of Applied Physics*, 38(9R):5262, 1999.

[31] C. Scharwitz, M. Böke, and J. Winter. Optimised plasma absorption probe for the electron density determination in reactive plasmas. *Plasma Processes and Polymers*, 6(1):76–85, 2009.

[32] B. Li, H. Li, Z. Chen, J. Xie, G. Feng, and W. Liu. Experimental and simulational studies on the theoretical model of the plasma absorption probe. *Plasma Sources Science and Technology*, 12(5):513, 2010.

[33] K. Takayama, H. Ikegami, and S. Miyazaki. Plasma resonance in a radio-frequency probe. *Physical Review Letters*, 5(6):238, 1960.

[34] A. M. Messiaen and P. E. Vandenplas. High-frequency dielectric resonance probe for the measurement of plasma densities. *Journal of Applied Physics*, 37(4):1718–1724, 1966.

[35] J. A. Waletzko and G. Bekefi. Rf admittance measurements of a slotted-sphere antenna immersed in a plasma. *Radio Science*, 2(5):489–493, 1967.

[36] N. Vernet, R. Manning, and J. L. Steinberg. The impedance of a dipole antenna in the ionosphere: 1. experimental study. *Radio Science*, 10(5):517–527, 1975.

[37] J. H. Kim, D. J. Seong, J. Y. Lim, and K. H. Chung. Plasma frequency measurements for absolute plasma density by means of wave cutoff method. *Applied physics letters*, 83(23):4725–4727, 2003.

[38] S. Dine, J. P. Booth, G. A. Curley, C. S. Corr, J. Jolly, and J. Guillon. A novel technique for plasma density measurement using surface-wave transmission spectra. *Plasma Sources Science and Technology*, 14(4):777, 2005.

[39] J. A. Fejer. Interaction of an antenna with a hot plasma and the theory of resonance probes. *Radio Science*, D68:1171–1176, 1964.

[40] R. S. Harp. The behavior of the resonance probe in a plasma. *Applied Physics Letters*, 4(11):186–188, 1964.

[41] R. S. Harp and F. W. Crawford. Characteristics of the plasma resonance probe. *Journal of Applied Physics*, 35(12):3436–3446, 1964.

[42] T. Dote and T. Ichimiya. Characteristics of resonance probes. *Journal of Applied Physics*, 36(6):1866–1872, 1965.

[43] A. J. Cohen and G. Bekefi. Linear and nonlinear response of a plasma sheath to radio frequency potentials. *The Physics of Fluids*, 14(7):1512–1524, 1971.

[44] J. Tarstrup and W. J. Heikkila. The impedance characteristic of a spherical probe in an isotropic plasma. *Radio Science*, 7(4):493–502, 1972.

[45] M. Lapke, T. Mussenbrock, R. P. Brinkmann, C. Scharwitz, M. Böke, and J. Winter. Modeling and simulation of the plasma absorption probe. *Applied Physics Letters*, 90(12):121502, 2007.

[46] C. Scharwitz, M. Böke, S. H. Hong, and J. Winter. Experimental characterisation of the plasma absorption probe. *Plasma Processes and Polymers*, 4(6):605–611, 2007.

[47] M. Lapke, J. Oberrath, C. Schulz, R. Storch, T. Styrnoll, C. Zietz, P. Awakowicz, R. P. Brinkmann, T. Musch, T. Mussenbrock, and I. Rolfes. The multipole resonance probe: characterization of a prototype. *Plasma Sources Science and Technology*, 20(4):042001, 2011.

[48] C. Schulz, M. Lapke, J. Oberrath, R. Storch, T. Styrmoll, C. Zietz, P. Awakowicz, R. P. Brinkmann, T. Musch, T. Mussenbrock, and I. Rolfes. The multipole resonance probe: Realization of an optimized radio-frequency plasma probe based on active plasma resonance spectroscopy. In *IEEE Middle East Conference on Antennas and Propagation (MECAP 2010)*, pages 1–5. IEEE, 2010.

[49] M. Lapke, C. Schulz, J. Oberrath, R. Storch, T. Styrnol, P. Awakowicz, R. P. Brinkmann, T. Musch, T. Mussenbrock, and I. Rolfes. Usage of electromagnetic modeling of the multipole resonance probe. In *2011 International Conference on Phenomena in Ionized Gases*. Citeseer, 2011.

[50] C. Schulz and I. Rolfes. Investigation of interactions between plasmas and rf-diagnostics: Challenges of complex 3d-electromagnetic field simulations. In *2013 IEEE Antennas and Propagation Society International Symposium (APSURSI)*, pages 2181–2182. IEEE, 2013.

[51] C. Schulz, I. Rolfes, T. Styrnoll, P. Awakowicz, M. Lapke, J. Oberrath, T. Mussenbrock, R. P. Brinkmann, R. Storch, and T. Musch. The multipole resonance probe: Investigation of an active plasma resonance probe using 3d-electromagnetic field simulations. In *2012 42nd European Microwave Conference*, pages 566–569. IEEE, 2012.

[52] M. Lapke, J. Oberrath, T. Mussenbrock, and R. P. Brinkmann. Active plasma resonance spectroscopy: a functional analytic description. *Plasma Sources Science and Technology*, 22(2):025005, 2013.

[53] T. Styrnoll, S. Bienholz, M. Lapke, and P. Awakowicz. Study on electrostatic and electromagnetic probes operated in ceramic and metallic depositing plasmas. *Plasma Sources Science and Technology*, 23(2):025013, 2014.

[54] C. Schulz, T. Styrnoll, M. Lapke, J. Oberrath, R. Storch, P. Awakowicz, R. P. Brinkmann, T. Musch, T. Mussenbrock, and I. Rolfes. A novel radio-frequency plasma probe for monitoring systems in dielectric deposition processes. In *2012 International Conference on Electromagnetics in Advanced Applications*, pages 728–731. IEEE, 2012.

[55] C. Schulz, T. Styrnoll, P. Awakowicz, and I. Rolfes. Supervision and control of medical sterilization processes utilizing the multipole resonance probe. In *2013 IEEE MTT-S International Microwave Workshop Series on RF and Wireless Technologies for Biomedical and Healthcare Applications (IMWS-BIO)*, pages 1–3. IEEE, 2013.

[56] D. Pohle, C. Schulz, I. Rolfes, M. Oberberg, P. Awakowicz, A. Serwa, and P. Uhlig. An advanced high-temperature stable multipole resonance probe for industry compatible plasma diagnostics. In *2018 11th German Microwave Conference (GeMiC)*, pages 235–238. IEEE, 2018.

[57] D. Pohle, C. Schulz, M. Oberberg, A. Serwa, P. Uhlig, P. Awakowicz, and I. Rolfes. Progression of the multipole resonance probe: Advanced plasma sensors based on ltcc-technology. In *2018 48th European Microwave Conference (EuMC)*, pages 239–242. IEEE, 2018.

[58] R. Buckley. A theory of resonance rectification. the response of a spherical plasma probe to alternating potentials. *Proceedings of the Royal Society of London. Series A. Mathematical and Physical Sciences*, 290(1421):186–219, 1966.

[59] J. Oberrath and R. P. Brinkmann. Active plasma resonance spectroscopy: a kinetic functional analytic description. *Plasma Sources Science and Technology*, 23(4):045006, 2014.

[60] J. Oberrath and R. P. Brinkmann. Influence of kinetic effects on the spectrum of a parallel electrode probe. *Plasma Sources Science and Technology*, 25(6):065020, 2016.

[61] J. Oberrath. Kinetic damping in the spectra of the spherical impedance probe. *Plasma Sources Science and Technology*, 27(4):045003, 2018.

[62] J. A. Bittencourt. *Fundamentals of Plasma Physics*. Springer Science & Business Media, 2004.

[63] D. G. Swanson. *Plasma kinetic theory*. Crc Press, 2008.

[64] R. Balescu. Transport processes in plasmas. 1988.

[65] I. L. Klimontovich. Kinetic theory of nonideal gas and nonideal plasma. *Moscow Izdatel Nauka*, 1975.

[66] L. D. Landau and E. M. Lifshitz. *Course of Theoretical Physics*. Elsevier, 2013.

[67] C. Cercignani. The boltzmann equation. In *The Boltzmann equation and its applications*, pages 40–103. Springer, 1988.

[68] R. Fitzpatrick. *Plasma Physics: An Introduction*. Crc Press, 2014.

[69] Thomas H Otway. Mathematical aspects of the cold plasma model. In *Perspectives in Mathematical Sciences*, pages 181–210. World Scientific, 2010.

[70] T. Mussenbrock, D. Ziegler, and R. P. Brinkmann. A nonlinear global model of a dual frequency capacitive discharge. *Physics of plasmas*, 13(8):083501, 2006.

[71] T. Mussenbrock and R. P. Brinkmann. Nonlinear electron resonance heating in capacitive radio frequency discharges. *Applied physics letters*, 88(15):151503, 2006.

[72] P. E. Vandenplas. Electron waves and resonances in bounded plasmas. *American Journal of Physics*, 39(2):236, 1971.

[73] Y. Itikawa. Cross sections for electron collisions with nitrogen molecules. *Journal of Physical and Chemical Reference Data*, 35(1):31–53, 2006.

[74] J. T. Gudmundsson. *Notes on the electron excitation rate coefficients for argon and oxygen discharge*. Raunvísindastofnun Háskólans, 2002.

[75] M. Lapke. *Analyse und Optimierung der Multipolresonanzsonde als industrietaugliches Plasmadiagnostiksystem*. Logos Verlag Berlin, 2011.

[76] J. Oberrath. *Modellierung und Analyse aktiver Plasmaresonanzspektroskopie mit funktionalanalytischen Methoden*. Logos Verlag Berlin, 2014.

[77] B. Li, H. Li, Z. Chen, J. Xie, and W. Liu. Dual-role plasma absorption probe to study the effects of sheath thickness on the measurement of electron density. *Journal of Physics D: Applied Physics*, 43(32):325203, 2010.

[78] L. Verlet. Computer "experiments" on classical fluids. i. thermodynamical properties of lennard-jones molecules. *Physical review*, 159(1):98, 1967.

[79] C. K. Birdsall and A. B. Langdon. *Plasma Physics via Computer Simulation*. CRC press, 2018.

[80] F. H. Harlow, M. Evans, and R. D. Richtmyer. *A machine calculation method for hydrodynamic problems*. Los Alamos Scientific Laboratory of the University of California, 1955.

[81] O. Buneman. Dissipation of currents in ionized media. *Physical Review*, 115(3):503, 1959.

[82] J. M. Dawson. One-dimensional plasma model. *The Physics of Fluids*, 5(4):445–459, 1962.

[83] J. M. Dawson. Particle simulation of plasmas. *Reviews of modern physics*, 55(2):403, 1983.

[84] M. A. Hellberg. A computer simulation of the plasma resonance probe. *Journal of Plasma Physics*, 2(3):395–435, 1968.

[85] A. C. Calder and J. G. Laframboise. Terminal properties of a spherical rf electrode in an isotropic vlasov plasma: Results of a computer simulation. *Radio science*, 20(4):989–999, 1985.

[86] R. W. Hockney and J. W. Eastwood. *Computer Simulation Using Particles*. CRC Press, 2021.

[87] N. Metropolis and S. Ulam. The monte carlo method. *Journal of the American statistical association*, 44(247):335–341, 1949.

[88] F. F. Chen. *Introduction to Plasma Physics and Controlled Fusion Volume 1: Plasma Physics*. Plenum Press, New York, 1984.

[89] J. D. Huba. *NRL Plasma Formulary*, volume 6790. Naval Research Laboratory, 1998.

[90] J. D. Callen. Fundamentals of plasma physics. *Online Book*, 2006.

[91] D. Tskhakaya, K. Matyash, R. Schneider, and F. Taccogna. The particle-in-cell method. *Contributions to Plasma Physics*, 47(8-9):563–594, 2007.

[92] C. K. Birdsall. Particle-in-cell charged-particle simulations, plus monte carlo collisions with neutral atoms, pic-mcc. *IEEE Transactions on plasma science*, 19(2):65–85, 1991.

[93] D. Lymberopoulos and Demetre J Economou. Two-dimensional self-consistent radio frequency plasma simulations relevant to the gaseous electronics conference rf reference cell. *Journal of research of the National Institute of Standards and Technology*, 100(4):473, 1995.

[94] H. Kim, F. Iza, S. Yang, M. Radmilović-Radjenović, and J. Lee. Particle and fluid simulations of low-temperature plasma discharges: benchmarks and kinetic effects. *Journal of Physics D: Applied Physics*, 38(19):R283, 2005.

[95] W. C. Swope, H. C. Andersen, P. H. Berens, and K. R. Wilson. A computer simulation method for the calculation of equilibrium constants for the formation of physical clusters of molecules: Application to small water clusters. *The Journal of chemical physics*, 76(1):637–649, 1982.

[96] R. Y. Rubinstein and D. P. Kroese. *Simulation and the Monte Carlo Method*. John Wiley & Sons, 2016.

[97] V. Vahedi and M. Surendra. A monte carlo collision model for the particle-in-cell method: applications to argon and oxygen discharges. *Computer Physics Communications*, 87(1-2):179–198, 1995.

[98] H. R. Skullerud. The stochastic computer simulation of ion motion in a gas subjected to a constant electric field. *Journal of Physics D: Applied Physics*, 1(11):1567, 1968.

[99] W. Feller. An introduction to probability theory and its applications. *1957*, 1967.

[100] T. Mussenbrock, T. Hemke, D. Ziegler, R. P. Brinkmann, and M. Klick. Skin effect in a small symmetrically driven capacitive discharge. *Plasma Sources Science and Technology*, 17(2):025018, 2008.

[101] J. D. Jackson. *Classical Electrodynamics, 3rd edition.* John Wiley & Sons, New York, 1999.

[102] C. L. Phillips, J. M. Parr, E. A. Riskin, and T. Prabhakar. *Signals, Systems, and Transforms.* Prentice Hall Upper Saddle River, 2003.

[103] K.-U. Riemann. Theoretical analysis of the electrode sheath in rf discharges. *Journal of applied physics*, 65(3):999–1004, 1989.

[104] C. Runge. Über die numerische auflösung von differentialgleichungen. *Mathematische Annalen*, 46(2):167–178, 1895.

[105] W. Kutta. Beitrag zur naherungsweisen integration totaler differentialgleichungen. *Z. Math. Phys.*, 46:435–453, 1901.

[106] R. P. Brinkmann. Beyond the step model: approximate expressions for the field in the plasma boundary sheath. *Journal of Applied Physics*, 102(9):093303, 2007.

[107] R. P. Brinkmann. From electron depletion to quasi-neutrality: The sheath–bulk transition in rf modulated discharges. *Journal of Physics D: Applied Physics*, 42(19):194009, 2009.

[108] V. A. Godyak, R. B. Piejak, and B. M. Alexandrovich. Electron energy distribution function measurements and plasma parameters in inductively coupled argon plasma. *Plasma Sources Science and Technology*, 11(4):525, 2002.

[109] T. Styrnoll. *Die Multipolresonanzsonde: Vom Demonstrator zur industrietauglichen Plasmadiagnostik.* PhD thesis, Ruhr-Universität Bochum, 2016.

[110] M. Fiebrandt, M. Oberberg, and P. Awakowicz. Comparison of langmuir probe and multipole resonance probe measurements in argon, hydrogen, nitrogen, and oxygen mixtures in a double icp discharge. *Journal of Applied Physics*, 122(1):013302, 2017.

[111] C. Schulz, I. Rolfes, M. Oberberg, and P. Awakowicz. Collision frequency determination of low-pressure plasmas based on rf-reflectometry. In *2016 46th European Microwave Conference (EuMC)*, pages 807–810. IEEE, 2016.

[112] P. Scheubert, U. Fantz, P. Awakowicz, and H. Paulin. Experimental and theoretical characterization of an inductively coupled plasma source. *Journal of Applied Physics*, 90(2):587–598, 2001.

[113] V. A. Godyak and V. I. Demidov. Probe measurements of electron-energy distributions in plasmas: what can we measure and how can we achieve reliable results? *Journal of Physics D: Applied Physics*, 44(23):233001, 2011.

[114] V. A. Godyak and B. M. Alexandrovich. Comparative analyses of plasma probe diagnostics techniques. *Journal of Applied Physics*, 118(23):233302, 2015.

A Acronyms

APRS   Active Plasma Resonance Spectroscopy

RF   Radio Frequency

LP   Langmuir Probe

MRP   Multipole Resonance Probe

PluTO   Plasma and Optical Technologies

IMRP   Ideal Multipole Resonance Probe

LTCC   Low Temperature Co-fired Ceramics

PIC   Particle-In-Cell

MCC   Monte Carlo Collision

SP   Super-Particle

RK   Runge-Kutta

ODE   Ordinary Differential Equation

EEDF   Electron Energy Distribution Function

EEPF   Electron Energy Probability Function

FWHM   Full Width at Half Maximum

B Physical Constants

Symbol	Unit	Numerical value	Description
ε_0	$\mathrm{A^2\,s^4\,kg^{-1}\,m^{-3}}$	$8.854187817 \times 10^{-12}$	Vacuum permittivity
μ_0	$\mathrm{m\,kg\,s^{-2}\,A^{-2}}$	$1.25663706 \times 10^{-6}$	Vacuum permeability
e	$\mathrm{A\,s}$	$1.602176565 \times 10^{-19}$	Elementary charge
m_e	kg	$9.10938291 \times 10^{-31}$	Electron mass
amu	kg	$1.660538921 \times 10^{-27}$	Atomic mass unit
m_i	kg	$6.63385291 \times 10^{-26}$	Ion mass of Argon
c	$\mathrm{m\,s^{-1}}$	299792458	Speed of light in vacuum
k_B	$\mathrm{kg\,m^2\,s^{-2}\,K^{-1}}$	$1.3806488 \times 10^{-23}$	Boltzmann constant

C List of Symbols

Symbol	Unit	Description
a	$\mathrm{m\,s^{-2}}$	Acceleration
$\boldsymbol{B}$	$\mathrm{T} = \mathrm{kg\,A^{-1}\,s^{-2}}$	Magnetic field
C	$\mathrm{F} = \mathrm{A^2\,s^4\,kg^{-1}\,m^{-2}}$	Capacitor
C_{die}	F	Dielectric capacitor
C_{she}	F	Sheath capacitor
C_{vac}	F	Vacuum capacitor
$\boldsymbol{D}$	A	Displacement current
d	m	Thickness of dielectric
$\boldsymbol{E}$	$\mathrm{V\,m^{-1}} = \mathrm{kg\,m\,A^{-1}\,s^{-3}}$	Electric field
$\boldsymbol{F}$	$\mathrm{N} = \mathrm{kg\,m\,s^{-2}}$	Force
f	$\mathrm{m^{-6}\,s^{-3}}$	Distribution function
f_{pe}	$\mathrm{Hz} = \mathrm{s^{-1}}$	Electron plasma frequency
f_{res}	$\mathrm{Hz} = \mathrm{s^{-1}}$	Resonance frequency
f_{EEDF}	$\mathrm{m^{-3}\,eV^{-1}}$	Electron energy distribution function
f_{EEPF}	$\mathrm{m^{-3}\,eV^{-3/2}}$	Electron energy probability function
f_{Maxwell}	$\mathrm{m^{-3}\,eV^{-1}}$	Maxwell velocity distribution function
f_{ϵ}	$\mathrm{m^{-3}\,eV^{-1}}$	Maxwell energy distribution function
H	$\mathrm{J} = \mathrm{kg\,m^2\,s^{-2}}$	Hamiltonian
$\boldsymbol{I}$	$\mathrm{A\,m^{-2}}$	Current
i	–	Imaginary unit
$\boldsymbol{j}$	$\mathrm{A\,m^{-2}}$	Current density
K	$\mathrm{m^3\,s^{-1}}$	Rate constant
L	$\mathrm{H} = \mathrm{kg\,m^2\,s^{-2}\,A^{-2}}$ or m	Inductor or characteristic length
L_{pl}	H	Inductor of plasma bulk
m	kg	Mass
m_{e}	kg	Electron mass
m_{i}	kg	Ion mass
N	–	Number of particles
N^{real}	–	Number of real particles
N^{s}	–	Number of super-particles
n	$\mathrm{m^{-3}}$	Density
n_{∞}	$\mathrm{m^{-3}}$	Quasi neutral density
n_{e}	$\mathrm{m^{-3}}$	Electron density
n_{i}	$\mathrm{m^{-3}}$	Ion density
n_{N}	$\mathrm{m^{-3}}$	Neutral particle density

P	$-$	Probability
$\hat{\mathrm{P}}$	$-$	Projection operator
$\underline{\underline{P}}$	$\mathrm{Pa} = \mathrm{kg\,m^{-1}\,s^{-2}}$	Stress tensor
p	$\mathrm{Pa} = \mathrm{kg\,m^{-1}\,s^{-2}}$	Pressure
p_{gas}	Pa	Pressure
$\boldsymbol{p}$	$\mathrm{kg\,m\,s^{-1}}$	Momentum
$\boldsymbol{Q}$	$\mathrm{kg\,s^{-3}}$	Energy flux density
Q	$\mathrm{A\,s}$ or $-$	Electric charge or quality factor
$\bar{Q}$	$\mathrm{A\,s}$ or $-$	Charge on the electrodes
$Q_{\mathcal{S}}$	$\mathrm{A\,s}$	Surface charge
q	$\mathrm{A\,s}$	Charge (positive or negative)
R	m or $\Omega = \mathrm{kg\,m^2\,A^{-2}\,s^{-3}}$	Radius of MRP or Resistance
R_{E}	m	Radius of electrodes
R_{f}	$-$	Random number $\in [0,1]$
R_{pl}	Ω	Resistance
$\boldsymbol{r}$	m	Position vector
S	$\mathrm{m^{-3}\,s^{-1}}$	Net change of particle density
T_{e}	eV	Electron temperature
T_{i}	eV	Ion temperature
T_{N}	K	Gas temperature
t	s	Time
U	$\mathrm{V} = \mathrm{kg\,m^2\,A^{-1}\,s^{-3}}$	Voltage
u	V	Voltage
$\bar{u}$	V	Symmetric applied voltage
$\boldsymbol{u}$	$\mathrm{m\,s^{-1}}$	Flow velocity
V	V or J	Voltage or potential energy
v_{th}	$\mathrm{m\,s^{-1}}$	Thermal velocity
$\boldsymbol{v}$	$\mathrm{m\,s^{-1}}$	Velocity
W	$-$	Numerical weight
$\boldsymbol{w}$	$\mathrm{m\,s^{-1}}$	Relative velocity
x, y, z	m	Cartesian coordinates
Y	$\mathrm{s^3\,A^2\,kg^{-1}\,m^{-2}}$ or $-$	Admittance or spherical harmonics
$\mathbb{Z}^{+}$	$-$	Positive integer
$\boldsymbol{\Gamma}$	$\mathrm{m^{-2}\,s^{-1}}$	Particle flux density
ΔQ	$\mathrm{A\,s}$	Charge difference of electrodes
Δt	s	Numerical time step
Δu	V	Asymmetric applied voltage
Δx	m	Numerical grid size
$\Delta \Psi$	$-$	Antisymmetric characteristic function
$\Delta \omega$	$\mathrm{s^{-1}}$	Half-width
δ	m or $-$	Sheath thickness or Kronecker delta
ϵ	J	Energy
ε_{p}	$\mathrm{A^2\,s^4\,kg^{-1}\,m^{-3}}$	Plasma permittivity
η	$-$	Resonance coefficient
Λ	$\mathrm{m^3}$	Debye volume

λ_{D}	m	Debye length
ν	s^{-1}	Collision frequency
ν_{col}	s^{-1}	Elastic collision frequency
ν_{ee}	s^{-1}	Electron-electron collision frequency
ν_{eff}	s^{-1}	Effective collision frequency
ν_{kin}	s^{-1}	Kinetic collision frequency
$\boldsymbol{\pi}$	$\mathrm{kg\,s^{-2}\,m^{-2}}$	Collision operator
ρ	m^{-1}	Charge density
σ	m^2	Cross section
σ_{eff}	m^2	Effective cross section
σ_{p}	$\mathrm{s^3\,A^2\,kg^{-1}\,m^{-3}}$	Plasma conductivity
$\sigma_{\mathcal{S}}$	m^{-2}	Surface charge density
Φ	V	Potential
$\Phi_{\mathcal{S}}$	V	Potential of surface charge
Ψ	$-$	Characteristic function
$\bar{\Psi}$	$-$	Symmetric characteristic function
ω	s^{-1}	Frequency
ω_{pe}	s^{-1}	Electron plasma frequency
ω_{pi}	s^{-1}	Ion plasma frequency
ω_{res}	s^{-1}	Resonance frequency